STABILITY AND CHANGE IN HUMAN CHARACTERISTICS

# STABILITY AND CHANGE

*Benjamin S. Bloom*

NEW YORK · LONDON · SYDNEY

# IN HUMAN CHARACTERISTICS

PROFESSOR OF EDUCATION, UNIVERSITY OF CHICAGO

John Wiley & Sons, Inc.

Library of Congress Catalog Card Number: 64-17133
Printed in the United States of America

*To Sophie*

# PREFACE

The entire import of this book may be summarized in three propositions.

1. The relation between parallel measurements over time is a function of the levels of development represented at the different times.
2. Change measurements are unrelated to initial measurements but they are highly related to the relevant environmental conditions in which the individuals have lived during the change period.
3. Variations in the environment have greatest quantitative effect on a characteristic at its most rapid period of change and least effect on the characteristic during the least rapid period of change.

The book could also be summarized in a single mathematical formula $I_2 = I_1 + f(E_{2-1})$, where $I$ represents quantitative measures of a characteristic at two points in time and $E$ represents the relevant environmental characteristics during the intervening period.

A third way of summarizing the entire work would be to draw a single graph such as the following:

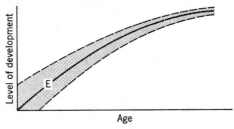

In this graph the single heavy line represents a typical developmental curve for a quantitative measure of some characteristic, and the shading represents the limits of variation that the environment can produce at different points in the development.

The book itself represents the fourth way of making explicit these same ideas and relationships. Of the four ways, the book is in many respects the least adequate and most cumbersome and awkward way of expressing these ideas. It attempts to relate various research findings to these ideas, attempts to state the "ideas" as they apply to many different characteristics, and, at least briefly, considers a number of theoretical as well as practical consequences of these ideas and findings.

It is unfortunate that the levels of communication in the behavioral sciences require these longer and more detailed treatments of ideas. It is to be hoped that the development of theory and method in the behavioral sciences will in the near future permit us to use the mathematical, graphic, and propositional forms of statements which can now be used in the physical sciences. It is likely that the more abstract as well as the more precise forms of communication will permit us to test the validity of ideas in far more adequate ways than is now possible with the longer and more easily misunderstood expansion of these same ideas.

BENJAMIN S. BLOOM

*Feburary 1964*

# ACKNOWLEDGMENTS

This work was begun during the year I was a Fellow at the Center for the Advanced Study of the Behavioral Sciences (1959–60). The freedom from the usual schedule and duties, the opportunity to explore a problem as deeply as possible, and the encouragement of the staff and other Fellows did much to help me get started on the problem of stability and change. The major outline of this work was completed at the Center.

It has, however, taken three additional years to complete the more detailed investigations and analyses of the relevant published works. In this, I have been assisted by Socrates Mamokus, Lee Shulman, and Richard Wolf. Dr. Wolf, in particular, has contributed to this work to such a level that I have difficulty at present in differentiating his work and ideas from my own.

Specific investigations directly related to this book were conducted by a number of my graduate students. In particular, I would mention the work of James Hicklin, Douglas Stone, Arlene Payne, Ravindra Dave, Neelakanthayya Hiremath, Marna Alexander, Allen Herzog, and Clarence Bradford.

The preparation of the manuscript, charts, and tables was a task that exhausted a number of secretaries. I am especially grateful to Mrs. Estell Jones, who took responsibility for the preparation of the final manuscript.

This work is in large measure based on the longitudinal studies conducted by many investigators in this country and abroad. Such studies require a type of patience and care unparalleled in other types of behavioral sciences research. The repeated measurement of a sample of persons for ten to thirty years requires a level of devotion and involvement that is very rare in the annals of research. I am heavily indebted to these many workers as well as to their subjects who cooperated in these long-term investigations.

Finally, I wish to express my appreciation for the encouragement and support for basic research given to my colleagues and myself by Francis S. Chase, Chairman of the Department of Education of the University of Chicago.

BENJAMIN S. BLOOM

# CONTENTS

# Chapter One

# INTRODUCTION TO STABILITY

# AND CHANGE IN HUMAN CHARACTERISTICS

THE PROBLEM

Recently I asked a number of my colleagues and graduate students to estimate the correlation between the height measurements of a sample of men at age 20 and age 40. Each person looked at me as though I had asked an absurd question. When pressed to answer this odd question, each declared that the correlation between height measurements at these two ages must be approximately +1.00. When asked why this was likely to be the case, they pointed out that almost all men have reached their full height growth by age 20 and that height is not likely to change during this period. When asked to estimate the correlation between height at age 10 and height at age 40, all were certain it would be relatively low but no one was able to give me an estimate in which he had any confidence. Several, however, did point out that the correlation between height at age 10 and age 20 would be the same as the correlation between height measured at age 10 and at age 40.

In this rather obvious example, it is evident that the magnitude of the correlation between the two sets of measurements is determined by the change or lack of change in the measurements of the individuals in the sample. When a human characteristic has reached its full development, the correlations between measurements of the characteristic are likely to be unity, except where there is some error in the measurements. Undoubtedly, there are some human characteristics which may continue to develop throughout the life of the individual. Also, there are human characteristics which reach approximately full development at age 20, age 15, or age 10, and perhaps some which may reach their full development at age 5 or even earlier. When we fully understand the course of the development of various human characteristics, we will better understand how to predict the development of the characteristic and its manifestations throughout the life of the individual. This understanding should also enable us to determine how and when change can be brought about as well as the consequences of change and stability.

Not all human characteristics are as visible and as obvious as stature. When are characteristics like honesty, empathy for others, aggression, etc., developed? Do these change throughout life? Can they be altered, and how, why, and when does this take place? When does general intelligence reach its full development and how and when can this characteristic be altered and to what extent? How early in life is the pattern of "grade getting" or academic achievement developed? Is it set by the end of elementary education? Perhaps a more fruitful way of posing the question is to ask not whether a particular pattern is stable or unstable, but to ask how stable the pattern is at various ages. What is the nature of the pattern of academic achievement and how does it develop quantitatively throughout life?

If we can describe the changing levels of stability of academic achievement, we can then determine the theoretical limits on the prediction of academic achievement over particular periods of time and at different age and grade levels. If we can further determine the effects of various factors and conditions on this stability, we may be able to determine how this pattern develops as well as how the pattern of academic achievement may be altered.

Although there is considerable theoretical and practical value in the analysis of the development and stabilization of school learning patterns, we are of the opinion that the identification of other "stable" human characteristics and the determination of the changing levels of stability of such characteristics may be of at least equal importance for our understanding of human growth and development.

This book, then, represents an attempt to identify "stable" characteristics, to describe the extent to which such characteristics are stabilized at various ages, and to determine the conditions under which this stability may be modified. Hopefully, this work will enable us to understand how such characteristics may be identified, explained, and, eventually, modified.

## SOME ATTRIBUTES OF STABLE CHARACTERISTICS

Empirically, a stable characteristic is one that is consistent from one point in time to another. If pressed, one might further delimit this by specifying the time intervals as one year or more and the minimum level of consistency as a correlation of $+.50$ or higher. Defined in this way, a stable characteristic may be one that is different quantitatively as well as qualitatively at the two time points if the change is predictable to some minimal degree. A good illustration of a stable characteristic from this point of view is height, which changes from birth to maturity, and yet height at maturity is highly predictable from height at ages 3 or 4. That is, the *relative* height positions of a sample of boys or girls are highly consistent from one age to another.

Although an empirical criterion of stability will enable us to identify "stable" characteristics and the extent to which they are stabilized at different ages, it is extremely important to be able to have a more theoretical method of identifying characteristics likely to be stable or unstable. Such an approach should enable us to identify stable characteristics on the basis of defined properties and then test the extent to which hypotheses are actually supported by empirical results. When, then, we find characteristics which satisfy our empirical but not our theoretical criteria, we may attempt to understand the discrepancy.

One aspect of a stable characteristic, which seems to be rather obvious and is clearly so by definition, is *nonreversibility*. By this, we mean that each change in the characteristic represents an increment added to the growth which has already taken place. We mean further that growth which has taken place at one time is not lost at a later time. Development is, from this point of view, additive and cumulative and a level of development once attained by the individual is not lost. Height very clearly meets this criterion since once a particular level of height has been attained, the individual retains it (at least until old age). Weight does not meet this criterion since the individual may control it and he may reduce or gain in weight at many different periods in his life. We suspect that learning of school subjects and the development of general intelligence are, to some extent, reversible. That is, the individual may have less of this characteristic at a later age than he did previously. However, much of the empirical evidence to be presented suggests that these characteristics are not readily reversed.

There is another aspect of reversibility of which we are not entirely certain. Can less than normal growth at one period of life be compensated for by greater growth at a later period?   In much of this work, we assume that the growth deficits of one period cannot be *fully* made up at a later period.   The data to be presented in succeeding chapters suggest that this is true of many characteristics if the deficit is incurred over a long period of time and if it occurs relatively early in the individual's development.   However, there is some evidence that deficits incurred over a short period of time may be *almost* fully recovered at a later period.   We suspect that a great deal of research will be needed to make clear the conditions under which deficits in growth may be recovered and the conditions under which such deficits cannot be recovered.

As we study the curves of growth of various characteristics, we note some in which growth begins very rapidly and then slows down (height).   Other characteristics develop at a relatively constant rate (weight), while still other characteristics grow very little in the early years and then accelerate at later ages (strength).   It would appear that characteristics which have negatively accelerated growth rates are most likely to be stable.   Such characteristics are also likely to be highly stabilized at an early age.   This is little more than saying that if little change takes place in a characteristic after a given age, the characteristic will be stable, and that stability is inversely related to change.

We find it difficult to make clear, but we believe that relatively superficial characteristics which the individual may develop in a short time (six months or less), characteristics which refer to highly conscious and easily controlled behaviors and mannerisms, and characteristics which are popular or dominant for a sizeable proportion of people at one stage in life (but which are not present at another stage) are not likely to be very stable.   Thus an interest in cowboys (movies, TV, and reading) at ages 5 to 11 seems to us to be a superficial interest which may completely disappear at a later stage of development.   The use of slang expressions, special types of dress, and the manners of adolescents appear to be temporary sets of characteristics which may not be evident before or after this stage of development.   In our search for stable characteristics, we are inclined to look for more pervasive and durable characteristics and to reject those characteristics which we believe to be temporary and transient.   We suspect that stable characteristics are more likely to be based on interactional processes, ways of relating to phenomena, life style, etc.   Immediate objects of attention, fads, and mannerisms are likely to be unstable characteristics unless they are viewed as symptomatic of persistent underlying needs and processes.   This is to say that basic mechanisms and processes are most likely to be stable, whereas symptoms and more superficial aspects of an individual's behavior are less likely to be stable. This may also be interpreted as the position that unconscious and deep-seated characteristics are difficult to change, whereas highly conscious and more nearly surface aspects of human behavior and personality may be more subject to change and are therefore less stable.

In the light of these criteria and defined aspects of stability, we might expect characteristics such as height, intelligence, academic achievement, generalized qualities of interest, and deep-seated personality characteristics to be stable and we should find empirical observations to be in support of this. In contrast, characteristics such as weight, interest in particular objects, specific learning (for example, typewriting, knot tying, arithmetic, and history), and more superficial personality characteristics and mannerisms are less likely to be stable either theoretically or empirically.

We must express a point of view in this work in that we are searching for stable characteristics and are less interested in either identifying unstable characteristics or in cataloging a large number of human characteristics. Furthermore, we shall emphasize those human characteristics which have been measured or quantitatively determined by techniques and instruments with acceptable levels of reliability and objectivity.

THE STUDY OF HUMAN CHARACTERISTICS OVER TIME

It is the central thesis of this book that the application of testing procedures as well as other highly focused systematic measurements and observations of individuals at different points in their history will, when properly summarized and analyzed, reveal fundamental facts about human growth and development. We can do little with retrospective accounts either by the subject or by observers if we are to trace the course and development of a particular characteristic from early childhood to maturity. What is needed is a series of independent observations at various points in the individual's career under such conditions that the observations at the different points in time are comparable and deal with the same human characteristic. From the facts obtained by such repeated measurement and observation, it should be possible to derive a set of generalizations about the development of particular human characteristics. These generalizations should then serve as the major propositions for a more comprehensive theory of human growth and development.

In education and the behavioral sciences, a great deal of attention has been paid to securing observations on an individual or a group at a particular point in time, followed by attempts to analyze and interrelate the various observations. When growth is studied, it is frequently done by securing observations or measurements on different age samples. Inferences about development are then made from the comparisons of the results on the different age samples. If growth is studied on the same sample of persons over time, it is rare that the investigator will follow the sample for more than a year or two.

We can learn very little about human growth, development, or even about specific human characteristics unless we make fuller use of the time dimension. Efforts to control or change human behavior by therapy, by education, or by other means will be inadequate and poorly understood until we can follow behavior over a longer time period. It is hoped that the emphasis on longitudinal research in this book will point to appropriate methods of attacking problems in research on human characteristics which have long eluded solution. One such problem is concerned with how much change can be produced in a particular characteristic at various points in the individual's career, and the conditions necessary for such changes. Other problems are concerned with the ways in which the development of particular characteristics can be made maximal or optimal, and the ways and the conditions under which these same characteristics may be distorted or developed only minimally.

## CONTRIBUTION OF TESTS AND MEASUREMENTS TO THE STUDY OF STABILITY AND CHANGE

Although it is possible to trace the development of examining and testing procedures to as early as the second century B.C. in China, as has been done by Creel (1960), much of what we now know as the testing movement began around the turn of this century. The work of Galton and Pearson on the measurement of individual differences and the use of statistics in studying such measurements emphasized the widespread individual differences on almost any characteristic that it was possible to measure. It was, however, the development of the intelligence test by Binet and Simon (1905) which gave momentum to this whole movement. Binet's work on a test of general intelligence in France was a major force in focusing attention on the possibility of measuring complex human characteristics with a considerable degree of precision. Binet's work on an individual test of intelligence was shortly followed by the development of group tests of intelligence as well as tests of many specific aptitudes. In addition, much work was devoted to constructing tests to measure school achievement, interests, attitudes, and personality characteristics.

The underlying methodology for all these instruments was the observation or measurement of an individual at a particular point in time under standard sets of conditions, with a standard set of problems and questions, and with carefully prescribed and standard procedures for appraising and summarizing the individual's responses. Central to the interpretation of test results was the collection of normative data which could be used as a basis for defining and describing the results for a particular indivudal. Of great significance in this testing movement is the very sharp focus on the specific test performance or behavior of an individual at a particular point in time without regard to other characteristics of the individual at that point of time or without regard to the conditions under which he had lived prior to this time. A specific measurement of an individual at a particular point in time can be utilized independently of evidence on the circumstances surrounding the individual. Although this presumed virtue of the testing movement has many advantages, it is likely that these measurements will take on far greater significance when they are placed in a larger context of other measurements as well as when they are seen in relation to the environmental circumstances under which the individual has lived or will live.

Along with the development of the standardized test have come other techniques for systematically observing and measuring human behavior. Among these are systematic observational and interview techniques in which the observer or interviewer notes specific and objective information which then can be judged and appraised by either the observer or others at a later time. Rating techniques have also been devised to judge and compare individuals according to specified characteristics. Much has been done to improve rating techniques in order to secure more comparable evidence and judgments from a number of different workers. Sociometric techniques have also been developed in which individuals may indicate their judgment of and relationship with other individuals by means of systematic procedures. Other, more projective techniques have been used to get at subtle and less conscious aspects of human personality.

Although this is not the place to conduct an elaborate defense of these instruments, the major point to be made is that these instruments are useful for the systematic gathering of evidence, and when used at different points in the individual's history, the evidence can be studied for both consistency and lack of consistency over time. One virtue of these testing instruments, as well as the allied techniques mentioned, is that they can be applied at a particular point in time in the study of an individual and can be reported without reference to what was known about the individual at another point of time, or what was known about other characteristics of the individual. The restrictions of the test or observational sample to particular characteristics as well as to a particular way of securing the evidence make this material of great value in the comparisons from one point in time to another. If we are trying to determine whether a particular characteristic which appears at ages 18 or 20 was present to a degree at ages 3 or 4, it is necessary to be sure that we are engaging in similar operational definitions of the characteristic, and that we have used similar ways of gathering and treating the evidence at these two points in time. If at one point in time one sort of evidence is used, such as an observation, while at another point the evidence is based on a test, we may not be certain that these two measurements are equivalent. The lack of consistency in the two sets of data may very well arise from the fact that the evidence being secured is very different at these two different points in time. Although it may appear to the reader that the restriction of evidence to a particular characteristic may be so narrow as to miss the more complex and important human characteristics, still it is only by this very narrow focus that we can determine the consistency or lack of consistency of a particular characteristic over time.

Another virtue of these instruments is that it is possible to make them highly objective or at least to know the degree of objectivity present in the appraisal of the evidence. This makes it possible to study individuals over many years and to reduce, or at least control, errors arising from the use of different observers or testers. For some of the instruments that we will refer to in this work, the measurements are highly objective and the method of summarizing and scoring is similar over a period of 15 to 30 years. Where the degree of objectivity of the instrument is much lower, this fact can be taken into consideration in appraising the evidence.

Another characteristic of these instruments is their known reliability. Although none of these instruments is perfectly reliable, it is possible to know how consistent the results are likely to be over short time intervals by the use of reliability estimates. Throughout this work, we will attempt to limit our study to instruments of relatively high reliability (reliabilities of .85 or higher), although from time to time we will make use of measurements where the reliability is less than would otherwise be desirable. The level of reliability will be taken into consideration in our analysis of the evidence.

In summing up, then, the development of the testing movement makes it possible for us to determine, with some precision, the extent to which a particular characteristic changes over time. The comparison of measurements at different times in an individual's career is possible because these instruments enable one to focus on the same kind of evidence at different times, and because the instruments permit us to control and limit the amount of error arising from the use of different testers and observers, from the use of different instruments, and from other errors of measurement. It then becomes possible to measure a particular characteristic in an individual at age 3, at age 10, at age 18, and even at age 28, and have some assurance that a similar set of characteristics is being investigated at these widely different points in time on the same individual. Furthermore, the availability of normative data enables us to look at the results on an individual in relation to results obtained on a defined sample and to draw meaning from the shifts in the individual's placement as compared with the normative data. Finally, these instruments are valuable for longitudinal comparisons primarily because the operations in which both the observer and the subject engage are well defined. Thus in testing and other equally systematic data-gathering procedures, we have powerful techniques for detecting similarity and consistency of characteristics over time. When used in appropriate research designs they may be used to determine whether and under what conditions these characteristics change.

NEED FOR ENVIRONMENTAL MEASURES

Much of the effort of test and measurement specialists has been devoted to the measurement and study of individual differences. "How do individuals vary?" How can these variations be measured with precision?" These have been the fundamental questions posed in this field.

Although much research has been done to explain and understand the sources of this variation, relatively little has been done to measure the environments with which the individuals interact. In the opinion of this writer, much of what has been termed individual variation may be explained in terms of environmental variation. It is likely that individuals who grow up in a home that emphasizes precision in the use of language and in which a rich and varied form of the language is used will develop larger vocabularies and better use of the language than will individuals who grow up in a home where an impoverished and crude use of the language predominates.   To test this hypothesis, we must have not only a measure of the individual's vocabulary but also a measure of the language used in the home.   It is the latter that we lack.   Although it is possible to do research in which we assume that the amount of education the parents have had is a useful indirect indicator of the language used in the home, this assumption may not always hold true.

It is this dearth of direct measures of the environment that restricts our fuller understanding of the development of particular human characteristics.   If we hypothesize a relation between growth in a characteristic and relevant aspects of the environment, we must test such hypotheses by appropriate experimental designs and with longitudinal measures of both the characteristic and the environment. If we hypothesize that a major change in the environment will bring about corresponding changes in the characteristic, then again we will need measurements of the environment and its changes as well as measurements of the human characteristic and its changes.

Throughout this work, we will refer to various indices of the environment; but for the most part, these are relatively indirect indicators of the qualities of the environment.   As a result we will frequently be able to consider only the most extreme differences in the environment since it is only with respect to extremes that we can be certain that our indirect indicators are really differentiating relevant aspects of the environment.

It is to be hoped that this work will be useful in stimulating new research on the measurement of environments.   Such research should enable us to more fully understand stability and change in human characteristics.   In Chapter 6 we will attempt to summarize some of our present understanding of the effects of varied environments on the development of selected characteristics.   The need for more precise environmental measures will also be highlighted in that chapter.

## EVIDENCE AVAILABLE IN LONGITUDINAL STUDIES

Most of the evidence to be used in this work has been derived from the many longitudinal studies reported in this country and abroad during the last 50 years.   These studies share a common methodology in that a selected sample of individuals has been measured or observed at two or more points in time. In each study the measurements were made independently in that the worker making one set of measurements made the measurements and records without knowledge of the results of the previous measurements. In some studies rather elaborate precautions were taken to insure that the measurements were as independent as possible.

Most of the longitudinal studies referred to in the present work made use of tests or measuring instruments in which a highly focused procedure was applied at given points in time to sample particular characteristics. Thus a test of intelligence is applied to secure a sample of the individual's performance on a set of questions. The test may take about an hour to administer, and it is that hour of questions and answers which constitute the sample on which a particular score or measurement is based. The point to be made is that the longitudinal studies to be considered in this work are based on repeated and highly focused measurements, each of which is restricted to a limited sample of the individual's characteristics or behavior under standard conditions.

We find four kinds of longitudinal studies or data collections which are of value for the problems and questions with which this book attempts to deal.

## Major Longitudinal Studies

There are approximately eight major longitudinal studies that have attempted to secure a large variety of measurements on a well-defined sample. These studies have followed a particular sample for 10 years or more. Usually the sample has been confined to a group of individuals living within a specified geographical area, and losses in the sample have resulted from departures of individuals from the geographical area, lack of willingness to cooperate further, or death. Some of the major studies to which we refer are the Iowa Studies (Baldwin, 1921), the Harvard Growth Study (Dearborn and Rothney, 1941), the Chicago Study (Freeman and Flory, 1937), the Berkeley Growth Study (Jones and Bayley, 1941), the Brush Foundation Study (Ebert and Simmons, 1943), the Fels Institute Study (Sontag, Baker, and Nelson, 1958), the California Growth Study (Macfarlane, 1938), and the Michigan Study (Olson, 1955).

In most of these studies, the same individuals have been measured on many different characteristics. In some of the studies, the environmental conditions under which the individuals have been raised have been described or measured.

## Follow-up Studies of a Particular Sample on a Small Number of Variables

A number of studies have been reported in which a sample of individuals has been tested or observed at two or more times. In some of these studies the sample has been kept as complete as possible even though the individuals moved to different geographical locations. Terman's *Genetic Studies of Genius* (Terman and Oden, 1947) is a good illustration of this type of research in that a group identified very early in life has been followed for over 30 years even though the individuals moved to many parts of the United States. In other studies the data are restricted to an initial and retest measurement on one variable or a single instrument (Nelson, 1954, attitudes toward religion; Owens, 1953, performance on the Army Alpha). Others may include several variables (Kelly, 1955, attitudes, values, and emotional adjustment).

In most of the studies which fall into this general category, there is little information about the environments in which the individuals have lived during the period between initial and retests. For the most part we will attempt to use these studies as supportive and supplementary to the more complete studies.

## Experimental Studies on the Effect of Educational or Other Variables

Although these studies were primarily intended to determine the effect of some experimental condition, the fact that they include initial and retest procedures for both experimental and control conditions will permit us to make limited use of them as supportive of other more extended investigations of stability and change. Most frequently these studies follow a group of subjects for a year or more. The intention of the research worker was primarily to study the effects of particular educational or other conditions, and the fact that longitudinal data became a byproduct of the study may be regarded as a fortunate accident for our purposes.

Since these studies usually involve only a limited time span, we will consider them supplementary rather than central in this work.

## Longitudinal Data Collections

Many social institutions maintain records on their clientele over relatively long periods of time. An especially good collection of data which includes teachers' judgments, psychologists' and counselors' reports, and test scores is to be found in the schools. Since these records are kept primarily for administrative and official reasons rather than for longitudinal research purposes, they may not always lend themselves to quantitative analysis.

Under some conditions, however, these records may be the basis for excellent longitudinal studies. In a few special instances, these data on individuals may be supplemented by very useful data about the environmental conditions. We shall turn to school records from time to time to test particular hypotheses about growth, stability, and change.

## MAJOR QUESTIONS TO BE INVESTIGATED

The development of the testing movement over the past sixty years and the application of tests and other standard instruments to the collection of evidence on individuals do make it possible to ask some rather searching questions about human characteristics and the conditions under which they remain the same or change over a period of time. The fact that similar instruments have been repeatedly used in longitudinal studies over the last half century makes it possible to assemble results from many different investigations and to raise a series of research questions. Some of these questions have already been answered in part by other summaries of the same evidence. Here, however, we will try to pose the questions in a somewhat different framework.

Our first question in examining sets of longitudinal data on a particular characteristic is, "Do different investigations with different samples, in different geographical regions, and at different times yield similar results?" Here we are attempting to determine whether there is any order or lawfulness in the results obtained from a number of studies. Does a study done from 1905 to 1920 yield results very similar to a study done from 1920 to 1935? We shall attempt to determine whether the variation from one study to another is attributable to changes in the instruments, to the differences in the techniques used, to differences in the conditions under which data were gathered, or to differences in the variability of the different samples. When possible we shall attempt to describe the general trend emerging from the studies.

When we can identify a single study which is most complete and adequate with regard to data collected and procedures used and which yields a good approximation of the general trend, we shall subject this study to more detailed analysis. This detailed analysis will determine the interrelationships among the measurements at the various stages of growth. This analysis will study the effect of combining the different measurements to reduce errors of measurement and will contrast the results obtained from point-to-point measurements with those based on accumulated data to a particular age or grade level. For example, how does height at age 3 relate to that at age 18 as compared with the accumulation of height measurements between birth and age 3 correlated with height at age 18? Where relevant, the analysis of a single study will also contrast sex and other groupings of the subjects.

Another question for our analysis will focus on the relationship between longitudinal data and cross-sectional or normative data. Under what conditions can cross-sectional data yield results which are congruent with longitudinal data? Can the two approaches be used to supplement or complement each other? Can one approach be used to predict or estimate the results obtained by the other approach?

We are interested not only in describing the development of particular characteristics but also in determining the conditions under which development may be altered. Specifically, an attempt will be made to determine the effects of differing environmental conditions on the growth and development of selected characteristics. How much effect can extreme environmental forces have on the development of a characteristic? Is the magnitude of these effects determined by the stage in the individual's development at which various environmental forces are introduced?

Closely related to this is the determination of the limits within which a characteristic may be altered by educational or other environmental forces. What are the limits within which growth in height may be influenced by extremes of environmental variation? In attacking this question, we shall make use of longitudinal studies where possible, but we shall also make use of other research which appears to be relevant—including studies of twins, normative studies done under varying conditions, as well as research on the effects of various experimental and control conditions. Although we are seeking the limits of environmental effects, it is more likely that we will be able to find the limits which have actually been observed or which may be inferred from existing studies, rather than the theoretical limits of environmental-individual interactions. Even so, we believe the attempt to locate observed limits will give an approximation to theoretical limits.

In any case, the designation of limits, whether observed or theoretical, should enable us to anticipate some of the results of future investigations in education and the behavioral sciences. Can we alter general intelligence? If so, how much change can be expected from different environmental or experimental conditions? How much alteration is likely to take place if these conditions are present for varying lengths of time at different stages in the developmental process? Can we alter problem-solving abilities? How much, under what conditions, etc.?

The answers to these questions can do much to help us in attacking some practical problems of education, child development, etc. In addition, the answers are especially important if education and the behavioral sciences are to move from a purely empirical view of human stability and change to a more theoretical viewpoint, where empirical findings can be explained by theoretical constructs, and observations can be related to a set of theoretical limits or values.

Finally, this work will attempt to relate the curves describing the development of each characteristic to the literature on the influence of early experiences and to the theoretical works on the development of human characteristics. Thus the work of Freud (1933), Erickson (1950), Horney (1936), and Fromm (1941) on the psychological and emotional development of the individual will be related to the findings on the growth and development of personality, attitudes and values, and interests. The work of Hebb (1949) and McClelland (1951) will be related to the patterns of growth in learning, educational achievement, and intelligence. The works of Elderton and Pearson (1915), Galton (1883), and Burt (1955) will be related to some of the nature-nurture problems posed by our data. Some of the work in comparative psychology, genetics, and animal breeding will be related to the findings of our research on the way in which the genetic potential of the organism is affected by environmental variation. Finally, some of the newer work on imprinting (Hess, 1959) and the effects of isolation and sensory deprivation (Solomon, 1961) will be related to the evidence on the early development of selected human characteristics.

It is to be hoped that this work will enable us to determine the consequences of various individual-environmental interactions, and to point up the way in which various human characteristics are set and then reinforced and stabilized by internal as well as external forces. Hopefully, this work will enable us to set some of the theoretical limits of human variation and to provide theoretical values against which various efforts to alter human characteristics by educational and other environmental forces may be appraised.

## REFERENCES

Baldwin, B. T., 1921.   The physical growth of children from birth to maturity.   *University of Iowa Studies in Child Welfare*, **1**, 1.

Binet, A., and Simon, T., 1905.   Méthodes nouvelles pour le diagnostic du niveau intellectual des anormaux.   *Année Psychol.*, **11**, 191–244.

Burt, C., 1955.   The evidence for the concept of intelligence.   *Brit. J. Ed. Psychol.*, **25**, 158–177.

Creel, H. G. (quoted in), 1960.   University of Chicago Reports, Chicago: Univ. of Chicago, Office of Public Relations, Vol. 11, No. 3.

Dearborn, W. F., and Rothney, J. W. M., 1941.   Predicting the child's development. Cambridge, Mass.: Sci.-Art Publishers.

Ebert, E., and Simmons, K., 1943.   The Brush Foundation study of child growth and development.   Washington: *Soc. Res. Child Develpm.*, **8**, No. 2.

Elderton, E. M., and Pearson, K., 1915.   The relative strength of nurture and nature. London: Cambridge Univ. Press.

Erikson, E. H., 1950.   Childhood and society.   New York: Norton.

Freeman, F. N., and Flory, C. D., 1937.   Growth in intellectual ability as measured by repeated tests.   *Monogr. Soc. Res. Child Develpm.*, **2**, No. 2.

Freud, S., 1933.   New introductory lectures on psychoanalysis.   New York: Garden City.

Fromm, E., 1941.   Escape from freedom.   New York: Farrar and Rinehart.

Galton, F., 1883.   Inquiries into human faculty and its development.   London: Macmillan.

Hebb, D. O., 1949.   The organization of behavior.   New York: Wiley.

Hess, E., 1959.   Imprinting.   *Science*, **130**, 133–141.

Horney, K., 1936.   The neurotic personality of our time.   New York: Norton.

Jones, H. E., and Bayley, N., 1941.   The Berkeley growth study.   *Child Develpm.*, **12**, 167–173.

Kelly, E. L., 1955.   Consistency of the adult personality.   *Amer. Psychologist*, **10**, 659–681.

Macfarlane, J. W., 1938.   Studies in guidance.   Washington: *Soc. Res. Child Develpm.*, **3**, No. 6.

McClelland, D. C. et al., 1951.   Personality.   New York: William Sloane Associates.

Nelson, E. N. P., 1954.   Persistence of attitudes of college students fourteen years later.   *Psychol. Monogr.* No. 373.

Olson, W. C., 1955.   Child development.   Boston: Heath.

Owens, W. A., 1953.   Age and mental abilities.   *Genet. Psychol. Monogr.*, **48**, 3–54.

Solomon, P., et al., 1961.   Sensory deprivation.   Cambridge, Mass.: Harvard Univ. Press.

Sontag, L. W., Baker, C. T., and Nelson, V. L., 1958.   Mental growth and personality development: a longitudinal study.   *Monogr. Soc. Res. Child Develpm.* **23**, 1–143.

Terman, L. M., and Oden, M. H., 1947.   Genetic studies of genius *IV*.   The gifted child grows up.   Stanford: Stanford Univ. Press.

# Chapter Two

# PHYSICAL CHARACTERISTICS

Although much of the longitudinal research was intended to further our understanding of the mental, educational, emotional, and social growth of the child, many of these studies have also included measurements of physical development.

It is probably safe to surmise that quite frequently data on physical growth were included because of the relative ease with which the measurements could be obtained. In many of the early longitudinal studies, the investigators were not entirely certain which data would turn out to be meaningful or needed and which would prove to be relatively valueless. Under such conditions, the workers tended to gather more rather than less data, especially when the data could be obtained and recorded with a minimum of extra cost and time.

A number of pioneers in longitudinal research, Jones (1948), Baldwin (1921), Olson (1955), and Dearborn and Rothney (1941), suspected that physical development was likely to be related to psychological development. Some hypothesized a very direct relation between physical change and psychological characteristics. Others were looking for some relation between the individual's *perception* of his own physical growth and development in relation to that of his peers and the consequent *effects* of this perception on his own social and psychological development. Thus an adolescent girl who was relatively immature in physical characteristics (in relation to her peers) might have special difficulties in her social life, and in turn her distress about this immaturity might have considerable effect on her emotional adjustment, attitudes, and interests.

Our interest in physical characteristics in this work is less for their developmental features per se than for their utility as models for the analysis of other types of growth.  The analysis of the available data on the stability and development of physical characteristics should help us to find techniques for analyzing other types of development.  The relative simplicity, at least from a measurement point of view, of data on physical development should enable us to become somewhat clearer about the more abstract and indirect types of measurement data available in other areas.  We shall use the longitudinal evidence on physical characteristics as a type of Occam's razor to help us maintain a simplicity of analysis and explanation as we move to the more complex types of growth data.


HEIGHT

We shall at a later point in this chapter bring together some of the data on the stability of other physical characteristics.  However, we shall devote our major attention in this chapter to the analysis of data on height.  We would refer to it as standing height, except for the fact that in several of the studies the measurements have been obtained by having the subject lie on a mat while his body length is measured.

   Height is a relatively simple physical characteristic to measure.  In most of the studies, height has been measured by having the subject stand erect barefooted while a bar attached to a centimeter scale is pressed down until it touches his head.  Typically, this is done three times and the three readings averaged.  The reliability of such measurements is indicated by the average difference of about 3 millimeters in the height of a group of subjects on two independent sets of measurements made within a few days by different workers (Krogman, 1950).  The reliability of these measurements when carefully made is also demonstrated by correlations of +.96 to +.99 between measurements made a year apart (Tuddenham and Snyder, 1954, p. 212).

   These high reliability figures make it possible to use height measurements without corrections for unreliability.  In later chapters we shall refer to measurements which have reliability figures below +.90.  For such measurements, some consideration will be given to the types of relationships theoretically to be expected if the reliabilities were perfect—a level which is almost approximated in the measurement of height.  This is not to say that all measurements of height have as high reliabilities as those reported here.  Careless measurement techniques, poor cooperation from subjects, changes in procedures, poorly trained measurement workers, etc., all could produce very low reliabilities in the measurements of even as simple a characteristic as height (Krogman, 1950). It is noteworthy that none of the longitudinal studies used height measurements reported by parents. Each of these studies has used trained personnel and standard procedures for the measurements. Such procedures are the necessary prerequisites for meaningful measurement and for scientific validity in the study of the growth of any human characteristic.

We have found approximately a dozen longitudinal studies in which the height of children has been measured and correlational data reported. These studies have been published from 1915 to 1959 and include research done in Sweden, England, and Israel as well as in the United States. Because of differences in the physical development of boys and girls, most of the studies report separate data by sex. In addition, some studies, for example, Shuttleworth (1939) break down the data by ethnic origin because of differences in physical characteristics of different ethnic groups.

Most of the studies have attempted to secure physical and other measurements at or as near as possible to the subject's birthday. For some age periods, the subjects were measured at three or six month intervals. In some studies, when it was not possible to measure the individual on his birthday, some attempt was made to secure an approximation of the birthday measurement by interpolation. Most of the studies have reported data only for those cases where a complete series of measurements has been obtained. However, where occasional measurements were missing, interpolated values have been inserted. It is clear that not all the measurements reported have been obtained under identical conditions for all subjects.

CONSISTENCY OF DATA

One of our first questions in attempting to discern a pattern in these data is to what extent the various studies yield consistent values. In Chart 1 we have plotted the correlations for boys between height at each age and height at maturity (age 17 or older). Each line represents the data from one of the longitudinal studies.

It will be noted that, although the general pattern is the same, the Shuttleworth (1939) studies show a lower level of correlation than do the two other studies. The data for the girls in Chart 2 exhibit a much more consistent trend.

Each study represents a particular sample from a much more general population. The practical requirements in longitudinal research

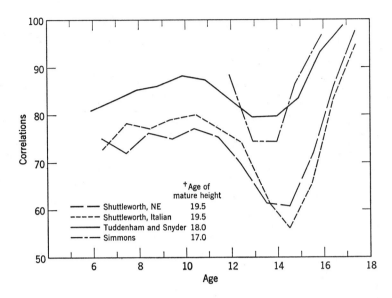

Chart 1.   *Correlations between Height at Each Age and Height at Maturity†—Males.*

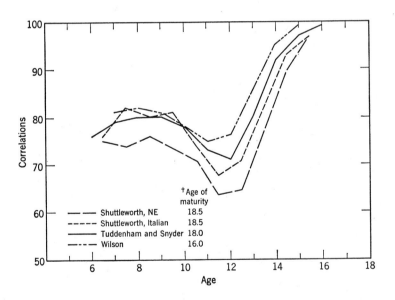

Chart 2.   *Correlations between Height at Each Age and Height at Maturity†—Females.*

for a group of persons who can be measured annually or more frequently over a period of years usually have imposed some limits
on the number and range of subjects included in each study. This
most frequently has meant that longitudinal studies are conducted
with subjects who continue to live in the vicinity of the headquarters
of the research office. As a result of selection influences, the sample
of subjects in each longitudinal study differs in both central tendency
and variability from a large representative population.

Kelley (1924) has developed a formula for estimating the effect of
changing the variability on the magnitude of the correlation between
two variables. It is possible to use this formula to show the effect of
changing the standard deviation from that found for the sample to
the standard deviation of a larger and more representative population.
This formula thus has the effect of showing the correlations to be
expected in a sample if it possessed the variability characteristics of
the larger population. It is unfortunate that no satisfactory formula
has been developed to correct for the standard deviations of both
variables in the sample (McNemar, 1949). Since the criterion variable
(height at maturity) enters into all the correlations for each curve
(Charts 1 and 2), it seemed soundest to apply the correction to the
criterion standard deviation. The representative population used is
the composite population tabulated from many different samples of
school age children by Martin (1953). His figures are based on a
national sample of over 200,000 children.

Charts 1a and 2a show the effect on the correlations of correcting
the standard deviations to the values in the composite population
developed by Martin (1953). It will be noted that the curves for
the girls are now almost indistinguishable from each other, whereas
the relationships for the boys are somewhat more consistent. We
believe the differences between the correlations for the Shuttleworth
samples and the other samples may be explained by the greater
heterogeneity of the environments represented in the Shuttleworth
data. The Shuttleworth samples are based on large groups in the
Boston area and include a very great range of socio-economic status
as well as ethnic background. The other studies tend to include more
homogeneous groups with regard to these parental characteristics.
We shall discuss this in greater detail later in the chapter.

The main generalization to be derived from these charts is that
results from the different longitudinal samples, when corrected to a
common terminal variability, reduce to a single pronounced trend.
Although the correlations, especially for the boys (Chart 1a), are different in magnitude, the same trends are clearly present in all the

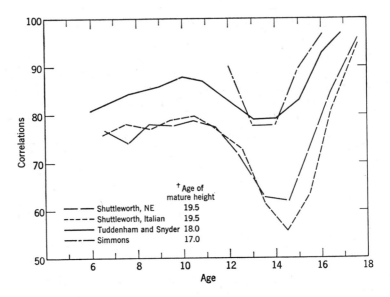

Chart 1a.   *Correlations between Height at Each Age and Height at Maturity* † *Corrected to a Common Terminal Variability—Males.*

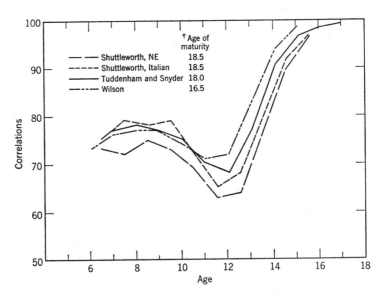

Chart 2a.   *Correlations between Height at Each Age and Height at Maturity* † *Corrected to a Common Terminal Variability—Females.*

studies.   The correlations for girls (Chart 2*a*) are so similar that it is difficult to distinguish the data of one study from those of another.

It should be remembered that these studies were reported from 1939 to 1959 and are thus representative of height measurements secured over a period of over 40 years of research (the study published in 1939 really began in 1920).   The lawful character of these data suggests that it is now possible to anticipate the interage relations in a study of height measurements, especially if one knows the standard deviation of the criterion age.   Thus one may turn the problem around from that of showing the consistency of the various studies completed to date to that of predicting the correlations to be obtained in new studies.

We have attempted to predict the correlations obtained in the Wilson study (1935) of girls' height and find that the Tuddenham and Snyder (1954) data which show the correlations of earlier height measurements with height at age 16 (the terminal age for Wilson) and the Wilson results are as follows:

| Ages | Tuddenham and Snyder | Wilson |
|------|------|------|
| 7–16 | .81 | .81 |
| 8–16 | .82 | .82 |
| 9–16 | .82 | .81 |
| 10–16 | .80 | .78 |
| 11–16 | .76 | .75 |
| 12–16 | .75 | .76 |
| 13–16 | .84 | .86 |
| 14–16 | .94 | .95 |
| 15–16 | .99 | .99 |

Thus the results of a study which has taken about 10 years to complete are quite accurately predicted by the general curve representing one or more previous studies.

FURTHER ANALYSIS OF ONE STUDY

The study of the *Physical Growth of California Boys and Girls from Birth to Eighteen Years* by Tuddenham and Snyder (1954) represents a most complete longitudinal study of physical growth.   This study, which was begun in 1928, followed a group of Berkeley children from birth to ages 18 to 20.   It includes 66 boys and 70 girls and is unusually complete in that it includes the raw data for each of the subjects as well as the statistical summaries.

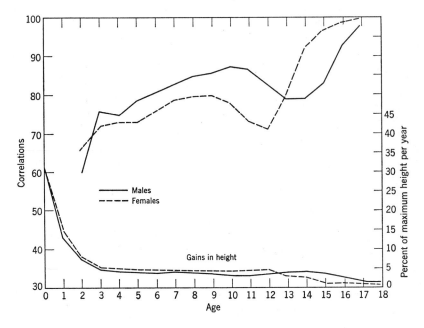

*Chart 3.  Correlations between Height at Each Age and Height at Maturity (Age 18) and Annual Gains Expressed as Percentage of Mature Height—Males and Females (Tuddenham and Snyder, 1954).*

It will be noted in Charts 1a and 2a that this study most closely approximates the general trend and is thus quite representative of height measurement interrelationships for the studies we have been able to bring together.

In Chart 3 we have shown the correlations between height at age 18 and height at each age from 2 to 17 for boys and girls. It will be noted that the correlations for boys increase from +.60 between ages 2 and 18 to +.88 between ages 10 and 18. The correlations then decrease to +.79 for ages 14 and 18, and again increase until they reach +.98 for ages 17 and 18. The results for the girls are very similar except that the highest correlation before adolescence for girls is +.80 between ages 8 and 18. It then declines to +.71 between ages 12 and 18, and again increases to +1.00 between 17 and 18.

We have attempted to explain the rise and fall in correlations in several ways. First, we have included in Chart 3 the gains in percent of maximum height for each age. Here, it will be noted that the correlations tend to be lowest for those years in which the gains in maximum height are greatest (that is, age 2 for boys and girls, ages

11 to 12 for girls, and ages 13 to 14 for boys). What appears to be happening is that during these very rapid growth periods some of the children are growing more rapidly than others and some children get their growth spurt earlier than do others. Thus in the adolescent period if a measurement is taken at age 14 for boys, one boy may have almost completed his adolescent growth spurt, another boy may at this age be in the middle of his spurt, and a third boy may only have just begun his spurt. The reported height measurement at age 14 does not (by itself) indicate where the individual is in his own physiological development.

It has already been pointed out by Wooley (1926), Shuttleworth (1939), and Greulich (1941), that the children who get their growth spurt earlier and who gain the most in these spurts tend to be the tallest at maturity. This is corroborated by the data in the Tuddenham and Snyder Study. Thus, in Chart 4, we have shown the relationship between the gains in different age periods and height at age 18. During the period of age 2 to age 13, the height gains of boys

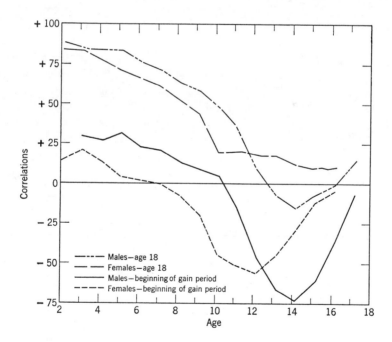

Chart 4. *Gains in Height from Each Age to Age* 18 *Correlated with Height at Age* 18 *and with Height at the Beginning of the Gain Period—Males and Females* (*Tuddenham and Snyder,* 1954).

are positively correlated with height at age 18. During the age period of 13 to 16 the gains are negatively correlated with height at age 18, whereas the gains from ages 16 to 18 are again positively correlated with height at age 18. The results for girls show low (but not negative) correlations between gains during the age period 10 to 17 and height at age 18. In general, gains in height prior to the adolescent growth spurt are positively related to mature height, while the gains toward the end of the adolescent spurt tend to be made by individuals who are not likely to be the tallest at maturity.

Another significant point which emerges from the study of gains is that the relation between gains and mature height varies from year to year.* In some years the relationship may be as high as +.40 while in other years it may be as low as −.40. Much the same may be said about the relations between height at age 2 and gains each year. If we average the correlations between height at age 2 and gains each year, the mean correlation for boys is +.06. Thus, on the average, the relationship between initial height (height at age 2) and the yearly gains in height is very close to zero.

In longitudinal studies of a characteristic like height the correlation between height at one age and height at another age may be thought of as a correlation between a *part* and a *whole* which includes the part. Thus, the correlation between height at age 2 and height at age 18 may be seen as the correlation between height at age 2 and *height at age 2 plus the gain in height from age 2 to age 18.* That is, the height at age 18 consists of the height at age 2 plus the gain in height from ages 2 to 18. This view of longitudinal data is most clearly true for characteristics which are additive in that what is attained by one age is not lost and is included in the measure of the characteristic at a later age. For longitudinal data, we may represent the relationship between two measurements in the following formulas

$$r_{X_1 X_2} = \frac{\sigma_{X_1} + r_{X_1(X_2 - X_1)}\sigma_{(X_2 - X_1)}}{\sigma_{X_2}}$$

or

$$r_{X_1 X_2} = \frac{\sigma_{X_1}}{\sigma_{X_2}} + \frac{r_{X_1(X_2 - X_1)}\sigma_{(X_2 - X_1)}}{\sigma_{X_2}}$$

* The relationships between annual gains and height at age 2 and age 18 *are not* shown in the form of a table or graph. Chart 4 shows the relationship between gains over an entire period (that is, ages 3 to 18, 5 to 18, etc.) and height at the beginning or end of the period.

where $X_1$ is the set of measurements at the initial time and $X_2$ is the set of measurements at the later time.

If the relationship between the first measurement and the difference (or gains) from the first to the second measurement is zero, the correlation will be equal to the ratio of the two standard deviations $(\sigma_{X_1}/\sigma_{X_2})$. Although all longitudinal data do not show this zero relationship between initial measurement and gains $(r_{X_1(X_2-X_1)})$, many of the studies do show relationships which approximate zero.

In any case, it is possible to see from these formulas that the Overlap Hypothesis (the overlap between two sets of measurements in a longitudinal series) may be broken down into two terms. One is the ratio of the two standard deviations and the other is the regression coefficient of the differences (or gains).

A second, and more general, approach to the Overlap Hypothesis is the interpretation of a correlation coefficient as the percentage of *elements* which are common to the two sets of measures involved. Kelley (1924, p. 190) proposes that if trait $X_1$ is determined by $N_c$ elements and trait $X_2$ is determined by these plus $N_b$ additional elements, then

$$r_{X_1X_2} = \sqrt{\frac{N_c}{N_c + N_b}}$$

or the square of the correlation $(r^2_{X_1X_2})$ is the proportion of elements determining $X_1$ which are involved in $X_2$. This approach to the Overlap Hypothesis is not limited to longitudinal series since it may be used to determine the percentage of elements common to two sets of measurements taken at the same time. What is meant by elements is left to the research worker. In our use of the Overlap Hypothesis we will interpret the elements as the units on our measurement scale that the two measurements $X_2$ and $X_1$ have in common.

A third approach to the Overlap Hypothesis will be the most general one. The correlation coefficient squared $(r^2_{X_1X_2})$ may be interpreted as the percentage of *variation* in one variable $(X_2)$ which is accounted for by another variable $(X_1)$. This approach makes no assumptions about the nature of the elements or the relationship between the difference and either of the variables. It may be applied to any set of variables and is not restricted to longitudinal data.

Anderson (1939) made use of the first two approaches. He suggested that for longitudinal data the correlation between the two measurements would equal the square root of the ratio of the two means $\sqrt{M_{X_1}/M_{X_2}}$ if the relationship between the initial scores and the gains was approximately zero. In other words, Anderson was hypothesizing that the correlations in longitudinal data are a direct function of the percent of the development at one age which has been attained at an earlier age. Anderson's formulation of the Overlap Hypothesis assumes an absolute scale with equal units and a defined zero. We shall make use of Anderson's interpretation of the Overlap Hypothesis throughout the book, where we believe his assumptions are likely to be met. The special value of Anderson's proposal is that it helps to relate cross-sectional and longitudinal data. Furthermore, it helps us to give additional meaning to the curves we shall draw to describe the relationships among the measurements.

In Table 2.1, we have shown the values to be expected from the Overlap Hypothesis in relation to the observed correlations for the height measurements for boys and girls in the California Study (Tuddenham and Snyder, 1954). We underlined all the expected values which are within .04 of the observed value—this is approximately at the .25 level of significance (a difference to be expected by

Table 2.1. Correlations between Height at Each Age and Height at Maturity (Age 18) and Expected Correlations from Overlap Hypothesis—Males and Females
Tuddenham and Snyder (1954)

| Age | 2 | 4 | 6 | 8 | 10 | 12 | 14 | 16 | 18 |
|---|---|---|---|---|---|---|---|---|---|
| 2 | | 83 92 | 77 87 | 75 82 | 72 79 | 67 76 | 61 73 | 62 71 | 60 70 |
| 4 | 80 92 | | 93 94 | 91 90 | 88 86 | 82 83 | 73 79 | 74 77 | 75 76 |
| 6 | 77 86 | 93 94 | | 97 95 | 95 91 | 88 88 | 77 84 | 79 82 | 81 81 |
| 8 | 75 82 | 90 89 | 97 95 | | 98 96 | 92 92 | 82 87 | 83 86 | 85 85 |
| 10 | 70 79 | 84 85 | 93 91 | 97 96 | | 96 96 | 86 92 | 88 90 | 88 89 |
| 12 | 62 75 | 77 82 | 88 87 | 91 91 | 96 96 | | 94 96 | 89 93 | 83 92 |
| 14 | 66 73 | 79 79 | 85 85 | 90 89 | 89 93 | 90 97 | | 92 97 | 79 96 |
| 16 | 66 73 | 74 79 | 77 84 | 82 88 | 80 92 | 75 97 | 94 99 | | 93 99 |
| 18 | 66 72 | 73 79 | 76 84 | 80 88 | 78 92 | 71 96 | 92 99 | 99 99 | |

The first correlation in each cell is the observed correlation. The second is the estimated correlation (based on the Overlap Hypothesis). The underlined estimated correlations fall within .04 of the observed correlations. Decimal points have been omitted. The correlations (estimated and observed) above the diagonal are for males and the correlations below the diagonal are for females.

chance 25% of the time).   It will be noted that for boys, almost all the expected values are within .04 of the observed values except for those at age 2 and at age 14.   For girls, almost all the expected and observed values from ages 4 to 14 are within .04 of each other.   At age 2 and after age 14, the majority of the differences is greater.

The disparity between the expected and observed values during the latter part of the adolescent period for both boys and girls seems to be accounted for by the negative relationships between height at these periods and the gains to maturity (see Chart 4).   However, the disparities between the expected and observed values for girls at age 18 are not satisfactorily explained.   We suspect that Anderson's version of the Overlap Hypothesis works less well in those periods where some individuals make small or zero gains, whereas others are making sizable gains—and this is true of the increments in girls' heights between ages 15 to 18.   This great differential in the rate of growth of individuals is true during the adolescent period for boys and girls and it is also true at about age 2—the end of a very rapid period of growth (conception to age 2).

If we neglect these periods of unusual and differential growth, the agreement between the values derived from Anderson's Overlap Hypothesis and the observed correlations is very high.   Thus we have two sets of data which are in very close agreement with each other. One is a set of normative data which can be secured from a cross-sectional study in which samples of different children are measured at each age, such as height norms for different ages.   The other data (correlations) must be derived from longitudinal studies in which the same children are repeatedly measured for periods as long as 18 to 20 years.   The agreement between the two suggests that a scale of development (the percent of mature height attained by earlier ages) can be related to a prediction scale (the relationships between height at maturity and height at earlier ages).   That is, the square of the correlation between any two ages is equivalent to the proportion that the height at one age is of the height at another age.

In the case of height measurement, we have both scales.   The measurements in terms of centimeters are expressed in absolute units which are equal and which are measured with a high degree of precision.   Later we shall refer to measurements of other human characteristics where only a set of relationships is available and we will wish to infer a developmental scale from these relationships.   Here, the agreement of both scales serves to demonstrate that one is, under certain conditions, the analogue of the other.

Referring again to Chart 3, we can now read the sets of relationships in terms of percent of mature development (using height at age 18 as mature development).  A male, on the average, has attained 30% of his mature height by birth (Baldwin, 1921), 54% by age 3, 86% by age 12, and 99% by age 17.  Put in other terms, 54% of boys' mature height is gained between conception and age 3, 32% of mature height is gained between ages 3 and 12, and 14% of mature height is gained between ages 12 and 18.  These figures demonstrate that the curve of height development is negatively accelerated.  The gains until age 3 are about 15% of mature development per year, the gains in the age period 3 to 12 are roughly 3% of mature development per year, and the gains in the age period 12 to 18 are about 2% of mature development per year.  Actually, the development in the period conception to age 3 changes from 30% from conception to birth to 5% in the age period 2 to 3, while in the age period 12 to 18 individuals may gain as much as 5% of mature height in a single year.  We will return to this notion of changing rate of growth later in this chapter when we will attempt to understand the effects of environmental influences on growth at different periods in the individual's development.

We will leave this topic at the moment with a reminder to the reader that an average male has attained 54% of his mature height by age 3 and that it is this very development that accounts for a correlation of +.76 between height at age 3 and height at age 18.  Thus the correlation may be thought of as a projection rather than a prediction. No great mystery is involved in the predictions of mature height. The correlations are determined by the proportion of mature growth. Put in other terms, the correlations in longitudinal studies of height growth are really not *predictions*, they are indications of the amount of growth which has *already* taken place.

## OTHER ANALYSES OF RELATIONSHIP

The foregoing represents an effort to relate height at two points, that is, correlation between height at age 3 and age 18.  However, data on height are available not only at age 3 but also at age 2 and at 3 month intervals before age 2.  This is the usual longitudinal situation.

With such data, we may now pose two questions.  What is the relationship between the evidence on height *up to a particular age* and mature height?  Are there some methods of treating the data which will yield higher levels of prediction than is available from the correlations between selected points?  This is the attempt to determine to what extent the particular characteristic (in this instance, height) is already present in the individual by a particular age.

Two concerns are present in this way of phrasing the problem. The first concern has to do with the reliability of the measurements. Any single observation may have error in it resulting from (1) errors in observing and recording or (2) difficulties in securing ideal measurements and cooperation from the subjects. Very little of the second type of error is likely to be present in a measurement of height or weight where the cooperation required of the subject is minimal. In more complex physical measurements requiring some cooperation from the subject such as strength, the error is likely to be greater. Such errors are likely to be very substantial in measurements which require good rapport and the full cooperation of the subject such as intelligence testing, problem solving and achievement testing, and attitude and value measurements. It is possible to reduce such errors of unreliability by simply summarizing the results of several measurements. Other errors in the measurements may be more difficult to eliminate.

A second concern is whether some way of combining the results of the measurements up to a particular age will increase the correlations with mature status as compared with the measurement at a single age. Will the developmental trends revealed in the series of measurements from ages 1 to 3 better predict the measurement at age 18 than will the single measurement at age 3? Will the evidence on physical growth from ages 1 to 3 improve the prediction of height at age 18 as compared with a single measurement at age 3?

Since the multiple correlation represents the best linear relationship between several predictors and a criterion, we have compared the multiple correlations for selected combinations of early measurements with the simple correlation when only one of the early measurements is used. In each case, (see Table 2.2) the last of the predictor measures is about as good a predictor of height at age 18 as is some combination of several predictor measures. This can be explained by the fact that the correlations among the predictors are so high that the combination of several yields no appreciable increase in the correlation with the criterion over the last in the series of predictor measures. It is evident that the correlations among the predictors are so high that the only point at which the use of the multiple correlation has advantages is in the adolescent period. Here the use of the multiple correlation results in a higher correlation with height at age 18 in contrast with the drop in the point-to-point correlation for the same period.

Table 2.2. *Some Multiple and Simple Correlations between Height at Age* 18 *for Boys and Selected Earlier Variables*

| Multiple Correlations | Simple Correlations |
|---|---|
| $R_{H_{18}::H_2:H_3} = .78$ | $r_{H_{18}:H_3} = .76$ |
| $R_{H_{18}::H_1:H_{1-2}} = .60$ | $r_{H_{18}:H_2} = .60$ |
| $R_{H_{18}::H_1:H_{21-36\,months}} = .75$ | $r_{H_{18}:H_3} = .76$ |
| $R_{H_{18}::H_{10}:H_{13}} = .84$ | $r_{H_{18}:H_{13}} = .79$ |
| | $r_{H_{18}:H_{10}} = .88$ |
| $R_{H_{18}::H_{10}:H_{14}} = .85$ | $r_{H_{18}:H_{14}} = .79$ |
| $R_{H_{18}::H_{10}:H_{15}} = .87$ | $r_{H_{18}:H_{15}} = .83$ |
| $R_{H_{18}::H_{13}:H_{15}} = .82$ | |

Another approach to the problem is to choose another measure such as socio-economic status and relate it to the height measurements at two ages to determine the effect of environment on the changes in height growth. Shuttleworth (1939) did report a slight effect of socio-economic status on height growth in both the Northern European and Italian groups. However, he made use of the entire sample in each group. We do not believe that height growth is likely to be influenced by small differences in socio-economic status since this is not a good approximation of the nutritional and other determiners of height in the environment.

In an attempt to test this hypothesis further, the Shuttleworth (1939) male sample was divided into a high economic group (parents in professional and semiprofessional occupations) and a low economic group (parents in unskilled occupations). It was reasoned that these two extreme economic groups would provide consistent though contrasting environments for physical growth. The average correlation between height at age 7 and height at age 17.5 for these two groups of males is +.90, as contrasted with the correlation of +.74 reported by Shuttleworth (1939) for all groups combined for these ages.

Further research is needed on this point, but it would appear that the prediction of future height is much more precise when a measure of the relevant environmental conditions is combined with an earlier measure of height. The stability of a characteristic such as height is determined by the extent to which the characteristic is developed and it is further determined by the environment in which the individual lives.

SOME DETERMINERS OF HEIGHT

There is considerable evidence that height development is responsive to three factors: inheritance, physical and chemical processes in the body, and disease and nutritional aspects in the environment. We shall discuss each of these briefly.

### Inheritance

At the beginning of the century, Elderton and Pearson (1915) determined the relation between the heights of parents and that of their offspring. This relationship, expressed as a correlation of approximately +.50, has been found repeatedly by other observers (Sanders, 1934). In general, the correlations rarely fall below +.50 and rarely rise above +.60.

Further evidence in support of the hereditary determination of height may be seen in the studies of twins and siblings. Burt (1955), Husén (1959), and Newman, Freeman, and Holzinger (1937) find a very close relationship (almost unity) between the heights of twins reared together as well as those reared apart (see Table 2.3). The

*Table 2.3. Correlations for Height and Weight of Twins and Siblings Reared Together and Reared Apart*

| | Height | | | Weight | | |
|---|---|---|---|---|---|---|
| | Burt and Conway | Newman, Freeman, and Holzinger | Husén | Burt and Conway | Newman, Freeman, and Holzinger | Husén |
| Identical Twins | | | | | | |
| Reared together | .96 | .98 | .89 | .93 | .97 | .81 |
| Reared apart | .95 | .97 | | .90 | .89 | |
| Nonidentical Twins | | | | | | |
| Reared together | .47 | .93 | .59 | .59 | .90 | .56 |
| Siblings | | | | | | |
| Reared together | .50 | | | .57 | | |
| Reared apart | .54 | | | .43 | | |
| Unrelated Children | | | | | | |
| Reared together | −.07 | | | +.24 | | |

relationship between fraternal twins or siblings reared together is lower than that for identical twins, but it is still about $+.50$ to $+.60$. The high relationship found by Newman, Freeman, and Holzinger for fraternal twins reared together must be questioned in the light of the findings by other research workers. Perhaps the difference is attributable to the difficulties involved in differentiating between identical and fraternal twins, with the most rigorous identification of identical twins being made by Newman, Freeman, and Holzinger. It is likely that some individuals classified as identical twins by the other workers would have been classified as fraternal twins by Newman, Freeman, and Holzinger.

The evidence on identical twins reared together as well as reared apart suggests that a very large amount of the variance in height (up to 90%) may be attributed to hereditary factors.

## Physical and Chemical Processes

The evidence for the cyclical nature of growth processes is very clear. This may be seen in the growth spurts identified by Rusch (1956) and in the differential growth processes of different parts of the body as described by Scammon (1930). Shuttleworth (1939) in his study of physical growth in relation to the menarche has made very clear that the age of puberty has a definite effect on the height growth of girls.

In the data supplied by Tuddenham and Snyder (1954) we find that the age of menarche has a very significant effect on growth. The age of menarche, for the 70 girls in this sample, has a correlation of $-.05$ with the height gains from ages 2 to 11, while the correlation is $+.70$ with the height gains from ages 11 to 18. This may also be seen in the fact that the ten girls who had the earliest menarche (ages 10.6 to 11.6) gained 12.4 centimeters in height from ages 11 to 18, whereas the ten girls who had the latest menarche (ages 14.0 to 15.8) gained 25.2 centimeters in the same period. This suggests that the height gains during adolescence are in part controlled by the onset of puberty and that the growth is greatest during this period if the menarche is delayed.

These results are in agreement with Gruelich (1941) and Stolz and Stolz (1951) who pointed up the physiological effects of the menarche in slowing up the pituitary secretions and in speeding up the gonad secretions. These changes in endocrine secretions have a major effect on height growth.

The point of all this is that height growth is clearly influenced by chemical processes in the body and that a partial determination of height growth can be achieved by influencing these processes.

## Disease and Nutritional Aspects

The many research studies on animals as well as humans establish the significance of amount and quality of nutrition on the skeletal development. Undoubtedly, major diseases at critical stages in the growth processes may also interfere with height development. We have chosen to regard nutrition and disease prevention as environmental determinants of height. We will not go into detail on the appropriate nutritional elements for height growth nor will we consider the diseases which may affect such growth. We will concern ourselves primarily with the major types of evidence that the environment can affect height growth.

Undoubtedly, human selection could have significant effects on height growth, but this is hardly a technique open to control by political or social institutions. Interference with chemical and physical processes in the body could clearly influence height growth, but as yet, this is likely to be done only in rare instances. It is not unlikely that such interference will become increasingly used and may even become as popular in the future as plastic surgery is at present for individuals with great concern about their physical appearance.

Man has gained increasing control over his environment and it is here that major strides have been taken throughout the world. It is in the shaping of the environment that we may expect the greatest impact on physical growth.

What is the effect of environment on height growth?

Keys, et al. (1950) reports height differences as great as 11% between children in Berlin orphanges before World War I and in 1919. If we accept, for the sake of further analysis, a value of 10% as the approximate difference in mature height between groups growing up under very extreme environmental conditions, ideal environment versus deprived environment, we may then ask what is the effect of extreme environmental conditions at different ages.

We may pose the question in several ways. If one group of children grow up under extremely favorable environmental conditions and another group grows up under extremely deprived environmental conditions, what should the difference between the two groups be at age 3, at age 11, and at age 18? If there is an extreme period of famine for three years, will there be any differences in the mature height of children who were born at the beginning of the famine as contrasted with children who were 6, 9, 12, or 15 years of age at the beginning of the famine? We may even put the question in practical terms. Will a given quantity of dietary supplement for children growing up under deprived conditions have a greater effect on height growth if it is used in the infancy period, in the primary school period, or in the secondary school period?

We may begin to answer this set of questions if, in addition to our figures on the normal development of height,* we make two assumptions.

1. Height growth is not reversible in that growth once attained is not lost except as a result of senility, surgery, or disease. This is an obvious truth about height development. Each gain in height is cumulative and mature height is the summation of all the gains in height from conception.

2. A somewhat more questionable assumption is that lack of growth or very limited growth in one period of development may not be *fully* recovered in a subsequent growth period. We hesitate to assert this to be true under all conditions, but it does seem likely that if the limitations on growth are severe and if they persist for long enough time, the deficiency cannot be fully compensated for at a later time. This assumption is supported by data on the growth of animals and it appears to be a major principle in accounting for data on human growth under extreme environmental conditions, especially during the critical growth years. We shall shortly present some evidence in support of this assumption.

Beginning then with the normal growth curve from conception to maturity, utilizing the 10% height difference at maturity, assuming nonreversibility from deficiency, we may hypothesize a series of height differences to be expected under normal, deprived, and abundant environmental conditions for height growth. These differences are indicated in a hypothetical table of the effect of environment at different age periods (Table 2.4).

In this table, we have assumed that at any age the effects of extreme environmental conditions may produce as much as a 5% difference from the normal growth expected. If, then, a child grows up until age 3 under very deprived conditions (as far as height is concerned),

---

* Both cross-sectional studies and longitudinal studies (Tuddenham and Snyder (1954), Scammon (1930), Greulich (1941), Simmons (1944), Stolz and Stolz (1951), Martin (1953) and Baldwin (1921) suggest that the average male gains about 54% of his mature height between conception and age 3, about 32% between ages 3 and 12 and about 14% between ages 12 and 18.

*Table 2.4. Hypothetical Effects of Different Environments on Height Growth of Males in Three Selected Age Periods*

| Age | Percent of Mature Height | Variation from Normal Growth | | | |
|---|---|---|---|---|---|
| | | Deprived | Normal | Abundant | Abundant-Deprived |
| Conception–3 | 54% | −2.7% | 0 | +2.7% | 5.4% |
| 3–12 | 32% | −1.6% | 0 | +1.6% | 3.2% |
| 12–18 | 14% | −0.7% | 0 | +0.7% | 1.4% |
| Total | | −5.0% | 0 | +5.0% | 10.0% |

this may have the effect of reducing his mature height by 5% of the base of 54%. This would mean a loss of almost 3% in potential mature height. If the deprived environment is encountered in the years 3 to 12, the 5% loss would mean a decrement of 1.6% in potential mature height, whereas a 5% loss in height growth at ages 12 to 18 would mean a 0.7% deficit in potential mature height. *The effect of deprivation or abundance is assumed* to be *greatest during the period of most rapid growth and least during the period of least rapid growth.*

What is the evidence in support of our values in Table 2.4?

## SOCIO-ECONOMIC DATA

There is considerable evidence, summarized by Sanders (1934), to demonstrate that at maturity the average difference in height between high and low socio-economic groups ranges from 8 to 20 centimeters, or an average of about 5% of mature height. However, these differences may reflect both hereditary and environmental influences. Furthermore, socio-economic status is not a very precise indicator of the quality of the environment for height growth.

Boas (1911) obtained height measurements on children born in Europe and in the United States of the *same* parents. He found that the heights of children born in this country in the first four years after migration were very similar to the heights of their siblings born in Europe. However, he found that children born in this country after their parents had been here four years or more were considerably taller than their siblings born abroad or their siblings born in this country shortly after the parents had migrated. The differences were of the order of 5%. The inference here is that although the children had similar hereditary characteristics, the United States environment and the improved economic status of the parents had a favorable effect on height growth. Many other studies have compared children of similar ethnic stocks born in the United States and in the country of origin (Tanner, 1955; Sanders, 1934). Usually the studies show a greater height for children raised in the United States, especially if the parents have a somewhat improved economic status in comparison with their countrymen abroad.

In Table 2.3 we have summarized some of the existing data on twins and siblings reared together and reared apart. It is clear that the correlation for height measurements between identical twins reared apart is almost as high as it is for identical twins reared together. However, this may be because of the lack of very great differences in the environments (for height growth) for each pair of children. Thus, of the 19 pairs of identical twins reared apart who were studied by Newman, Freeman, and Holzinger (1937), in no instance did the research workers report substantial differences in the physical care received by each pair. It would be very revealing if separated identical twins living under very dissimilar physical growth conditions could be studied.

FAMINE AND DEPRIVATION

Evidence from famine areas and devastation under war conditions, summarized by Sanders (1934), Keys (1950), and Tanner (1955) demonstrates the deficits from normal height suffered by different age groups. The evidence is very clear that in general children growing under deprived conditions are significantly shorter than children growing up under normal conditions in the same countries or cities. Keys (1950) summarizes the mean heights of boys in Berlin orphanages in 1919 as compared with boys in the orphanages before World War I. If we assume that the deprived conditions in Berlin were about three years in duration, then our application of the values developed in Table 2.4 would yield the following predicted values for the Berlin boys:

Age 3 (3 years of deprivation) equals 5% less than normal.

Age 12 (9 years of normal environment + 3 years of deprivation) equals 1.25% less than normal.

Age 14 (11 years of normal environment + 3 years of deprivation) equals 1.1% less than normal.

These figures may be contrasted with the percent deficiencies reported in Keys (1950, p. 992).

Age 3....11% less than normal.
Age 12....5.2% less than normal.
Age 14....7% less than normal.

The differences reported are considerably greater than our expected values.   It is possible that our estimates are too conservative, but it is also possible that some portion of the deficiency will be recovered by maturity.   In any case it is clear that at the ages measured the deficiencies were between 5% and 11% of the normal heights.

Dreizen and his co-workers (1953) studied the effects of nutritive failure on the height and weight of Alabama children since 1941. They report the heights of a group of children with nutritive failure

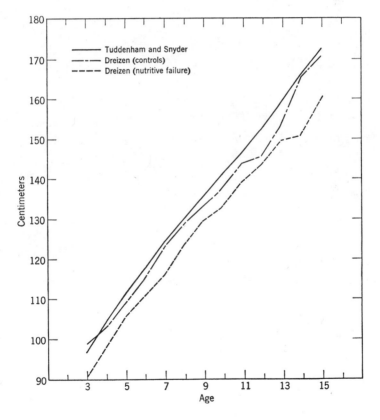

*Chart 5.   Height Growth under Different Environmental Conditions—Males.*

and the heights of an ethnically identical group of children without nutritive failure in the same geographic area. It will be noted in Charts 5 and 6 that the curves for the height of the nutritive failure groups become progressively differentiated away from the curve for the Tuddenham and Snyder samples, whereas the Alabama control groups without nutritive failure occupy an intermediate position. The differences between the nutritive failure groups and the Tuddenham and Snyder samples approach the percentages shown in Table 2.4 for deprived versus abundant environments. It is probably true that neither the Tuddenham and Snyder samples of Oakland children nor the Dreizen Alabama nutritive failure samples represent the theoretical extremes of abundance or deprivation.

Dreizen and his co-workers (1950) selected 41 children with nutritive failure and matched them with 41 control children also with nutritive failure. After observing both groups for 20 months, they gave milk supplements to the test group during the next 20 months but not to

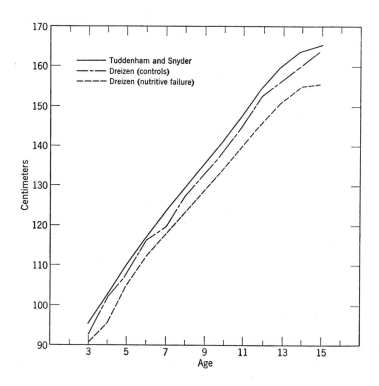

Chart 6.   *Height Growth under Different Environmental Conditions—Female.*

*Table 2.5. Mean Height (cm) and Weight Gains (kg) of
Children with Nutritive Failure during
Three Twenty-Month Periods*
*Dreizen et al.* (1950)

### Height Gains

| | Twenty Month Premilk Period | | Twenty Month Milk Period† | | Twenty Month Postmilk Period | |
|---|---|---|---|---|---|---|
| | Boys | Girls | Boys | Girls | Boys | Girls |
| Test Group‡ | | | | | | |
| 9 boys 10 girls | 8.18 | 7.91 | 11.77 | 14.60 | 7.33 | 7.84 |
| Control Group 18 boys 23 girls | 8.56 | 8.30 | 9.15 | 9.16 | 7.60 | 7.88 |

### Weight Gains

| | Twenty Month Premilk Period | | Twenty Month Milk Period† | | Twenty Month Postmilk Period | |
|---|---|---|---|---|---|---|
| | Boys | Girls | Boys | Girls | Boys | Girls |
| Test Group‡ | | | | | | |
| 9 boys 10 girls | 3.62 | 3.98 | 7.82 | 6.43 | 4.25 | 5.23 |
| Control Group 18 boys 23 girls | 4.15 | 4.88 | 4.42 | 5.05 | 4.35 | 4.26 |

† During the milk period the test group is significantly higher than the control group (.05 level or better).
‡ The test group, during the milk period, is significantly higher (.05 level or better) than the test group at other periods.

the control group, and then observed both groups for an additional 20 months during which no milk supplements were used. During the supplement period the test group made significantly greater height gains than they made during the pre- or post-supplement periods. During the supplement period the test group made significantly greater height gains than the control group (Table 2.5). It is of interest to note that the test group reverted back to their regular growth rate under nutritive failure when the milk supplements were withdrawn.

ANIMAL DATA

McCay and his co-workers (1939) underfed groups of rats for varying periods of time after weaning and compared their nose to anus length with a control group given a diet supplemented by cod liver oil. These results are summarized in Table 2.6. Although the periods do not correspond to our age periods for humans, it is clear that underfeeding in the rat does produce a deficit in skeletal development which is greater than we had anticipated for humans. McCay's (1935) work on another study (see Table 2.6) and two studies by Jackson (1936, 1937) are in general agreement with the findings indicated.

The research on domestic animals by McMeekan (1940) and Hammond (1959) demonstrates that skeletal development, musculature,

Table 2.6. *Mean Body Length (cm) of Experimental Rats Underfed for Different Periods of Time in Contrast with Control Rats*
McCay et al. (1935 and 1939)

| McCay et al. (1939) | Male | | Female | |
|---|---|---|---|---|
| Length of Underfeeding after Weaning† | Mean Body Length | Percent of Cod Liver Oil Control | Mean Body Length | Percent of Cod Liver Oil Control |
| 300 days | 21.2 | 95 | 18.6 | 92 |
| 500 days | 20.6 | 92 | 19.5 | 97 |
| 700 days | 20.0 | 90 | 17.7 | 88 |
| 1000 days | 18.3 | 82 | 18.0 | 90 |
| Control fed cod liver oil | 22.3 | 100 | 20.1 | 100 |
| McCay et al. (1935) | Male | Percent of Control | Female | Percent of Control |
| Restricted from time of weaning until death | 18.4 | 84 | 18.0 | 91 |
| Restricted from two weeks after weaning until death | 18.1 | 82.5 | 17.5 | 88 |
| Control (normal diet) | 21.9 | 100 | 19.8 | 100 |

† The experimental rats were fed normally after the underfeeding period.

shape, and amount of fat are highly controllable by varying the amount and quality of nutriments at different stages in the animal's development. The selective use of low and high planes of feeding at different growth stages can be used to produce an animal which conforms to the stock raiser's specifications. The research workers have determined the time tables for the development of skeleton, muscle, and fat for certain domestic animals and the feeding can be planned so as to maximize or minimize each of these characteristics.

GROWTH UNDER SPECIAL CONDITIONS

The most striking evidence should be found in situations where the parents have grown up under one set of environmental conditions while their children have grown up under a very different set of environmental conditions. Such a situation is most clearly approximated in some of the Kibbutz settlements in Israel. In some of these settlements, the parents were products of the ghettos of Central and Eastern Europe. In the Kibbutz settlements, the parents eat in one dining room while the children, until about age 15, eat in a separate dining room. The children are given a diet which is as nearly optimal as the nutritionist can devise. In many ways the diet for the children approximates what an international optimal diet might be, whereas the diet for the parents still reflects the particular ethnic food habits of the country of origin.

Although it has not been possible to collect the necessary measurements, the observations of the author and of a number of Israeli specialists, including physicians working in the Kibbutz settlements, are that the children at maturity tower over their parents. It is most common to see children and parents standing side by side with the male children being about a head taller than their fathers. Efforts are under way to collect the necessary precise data, but there is little doubt about the general order of magnitude of the height differentials of parents and children.

If we regard the parents as having grown up under conditions somewhere between deprived and normal, and the children as having grown up under conditions closely approximating an abundant physical growth environment, we have a crucial test of the hypothesis advanced previously as to the differential effects of extreme environments. The children have grown significantly taller than their parents.

It is likely that the correlation between the height of parents and children will be constant under a great many conditions. This relationship is indicative of hereditary influence on height growth.   However, the effect of the environment may be seen in the absolute differences between parents' and children's height.   If a group of children have grown up under environmental conditions which are similar to those of their parents, the correlation between parents' and children's height should be approximately +.50 (Elderton and Pearson, 1915), and the average size of the children will approximate that of their parents.   If, on the other hand, the parents have grown up under conditions markedly inferior to the conditions under which their children have been reared, the correlations should be about the same, but the children will be markedly taller than their parents.   If the parents have grown up under conditions markedly superior to the conditions under which their children have been raised, the correlation between parents and children's height should be the same, but the parents should be markedly taller than their children.

We would speculate that if one combines data in which one group of parents have grown up under conditions which are inferior to that of their children while another group of parents have grown up under environmental conditions markedly superior to the environmental conditions of their children, the correlation for the combined groups of parents and children will approximate zero and the average size of the combined group of children will be about the same as the average size of the combined group of parents.

What is suggested here is that the inheritance of characteristics can be studied only when it is possible to identify the environmental conditions under which both children and parents have been reared. This also leads to the inference that much of the early work on the relationship between parents and childrens' stature was in situations where parents and children grew up under similar conditions.   Historically, there were many situations in which diet and health conditions of parents and children were very similar.   As more becomes known about the control of diet and health conditions, it becomes increasingly possible to create environments for children which are markedly different from those under which their parents had been reared.   Under such conditions we may expect children to differ in height from their parents by an amount which is a function of the differences in environments.   All this does not minimize the importance of heredity.   It is merely a reminder that the hereditary potential for height transmitted from parents to children can be gauged from the parents' height only when it is clear that the parents have grown up under ideal circumstances for the development of stature.

OTHER PHYSICAL CHARACTERISTICS

Weight

Height appears to represent a highly stable characteristic and we will use it throughout this work as a reference point for other characteristics. Somewhat in contrast is weight which may go up as well as down over a period of time and which may be controlled almost at will by the individual. Several longitudinal studies by Baldwin (1921), Shuttleworth (1939), Simmons (1944) and Tuddenham and Snyder (1954) have included measures of weight as well as of height.

We have selected the Tuddenham and Snyder data for further analysis and plotted the correlations in Chart 7. The curve showing the correlations of weight at each age with height at age 18 is very similar for boys and girls. It will be noted that the developmental scale based on proportions of weight at age 18 reached by each age approximates the correlations except at the very early ages and at the

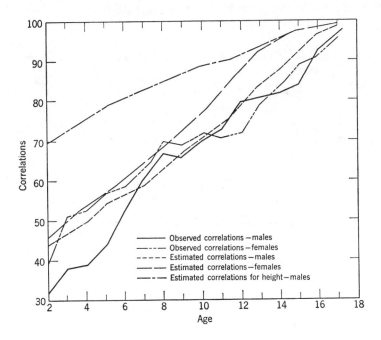

*Chart 7. Comparison of Observed Correlations for Weight with Correlations Estimated from Overlap Hypothesis—Males and Females (Tuddenham and Snyder, 1954).*

adolescent period for boys and girls. For contrast, we have also imposed the developmental scale for height on this chart. It is evident that the early growth rate for height is much greater than it is for weight. Thus the child at birth has reached the percent of mature development for height that will be reached at age 5 for weight. By age 2.5 the individual has reached approximately 50% of mature height, whereas this percent of development for weight is not reached until age 10.

We have not been interested in the question of the influences of the environment on weight. The relationships are so obvious and clear that few would hesitate to note the consequences of abundance or deprivation on weight. One could point to studies cited earlier for height by Dreizen (1953), McCay (1939), etc., showing the effect on both humans and animals of different levels of nutrition for varying periods of time. There is little doubt that weight growth deficits are more easily recovered, whereas height growth deficits are not likely to be fully made up if the deprivation takes place very early and if it is continued over an extended period of time.

The effect of environment is also clearly demonstrated in the studies of twins and siblings reared together and reared apart (see Table 2.3), Husén (1959), Burt (1955), Newman, Freeman, and Holzinger (1937). Here, the correlations between the weight of children reared together and reared apart are of approximately the same magnitude as those found for height.

## Strength

The data for strength measures at different ages are not as complete as the height and weight data. In Chart 8 we have plotted the correlations reported by Tuddenham and Snyder (1954) for males between strength at each age and strength at age 18. It will be noted that these correlations do not rise as rapidly as the correlations for height, although they are very similar to the values for weight. We have compared the observed correlations to those expected on the basis of the Overlap Hypothesis for strength. The fit is a very poor one, undoubtedly because our index of strength is composed of a combination of several measures of strength (right-hand grip, left-hand grip, chest pull, and chest thrust). A better approximation is found by using the estimates based on weight development, probably because of the general relationship between weight and strength.

It is evident that the measurement of strength requires more cooperation from the subjects than do measurements of height or

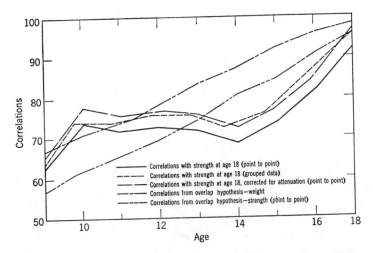

*Chart 8. Comparison of Observed Correlations (Point to Point and Grouped Data) for Strength with Correlations Estimated from Overlap Hypothesis—Males (Tuddenham and Synder, 1954).*

weight. We are of the opinion that strength cannot be measured as reliably as either height or weight. In order to demonstrate this we have computed the correlations in which two measures of strength at six month intervals have been averaged. These grouped data correlation values are somewhat higher than the correlations found for the point-to-point (age to age) correlations. The grouped data correlations are very close to the point-to-point correlations corrected for attenuation. Thus grouping of data has the effect of reducing the unreliability of the measures. We will show similar patterns in later chapters.

It is of interest for us to note that the measured strength of girls does not increase after age 13 while boys double their strength indices between ages 12 and 18. We suspect, as do Tuddenham and Snyder (1954), that this is partially because of the differences in cultural attitudes of boys and girls toward strength.

### Comparisons

We have drawn three scales (Chart 9) to indicate the timing of growth in height, weight, and strength for boys from the Tuddenham and Snyder data. Here it may be seen that the three characteristics develop at very different rates. The correlations reported below

*Percent of Mature Development*

| | 25 | 50 | 75 | 100 |
|---|---|---|---|---|
| Height | | | | |
| Age | Birth | 3 | 9 | |
| | \| | \| | \| | |
| Correlation with maturity | ? | .76 | .86 | |
| Weight | | | | |
| Age | 4 | 10 | 14 | |
| | \| | \| | \| | |
| Correlation with maturity | .39 | .70 | .82 | |
| Strength | | | | |
| Age | 9 | 12 | 16 | |
| | \| | \| | \| | |
| Correlation with maturity | .63 | .73 | .82 | |

*Chart 9. Development of Selected Characteristics of Males (Tuddenham and Snyder, 1954).*

each of these scales show the way in which the predictability (and stability) of human characteristics reflects the extent to which individuals have attained various levels of development.

Stability of physical characteristics is clearly determined by the rate of growth of each characteristic. The correlation coefficient between the measurements of the characteristic at given ages and at maturity serves not only as an index of the stability of the characteristic but also as an indicator of the percent of mature development attained by given ages.

SUMMARY

When the different longitudinal studies of height are compared, there is a great deal of consistency of relationships in the age-to-age measurements. The similarity of the pattern of correlations becomes even greater when appropriate statistical techniques are used to take into consideration the variability of each of the samples with respect to mature height. The consistency of the relationships suggests that the developmental aspects of height are strikingly similar from study to study even though the studies were done by different workers, with different samples, and at different times during the past half century. Thus, in spite of secular differences in the mean height at maturity and in spite of differences in the ethnic and socio-economic backgrounds of the samples studied, the patterns of relationship are very consistent. This finding for height, which will be demonstrated throughout this book for other characteristics, highlights the value of longitudinal studies for the analysis of the development of human characteristics. It is likely that a small number of carefully done longitudinal studies may be sufficient to establish some of the major developmental aspects of particular human characteristics.

The further analysis of one longitudinal study, Tuddenham and Snyder (1954), reveals the part-whole relationship which is true of many stable human characteristics. That is, the height developed to one age is included in the measurement of height at a later age. Thus the height developed to age 5 is included as a part of the height of the individual at age 10 or at maturity. The relationship between the earlier and the later measurement may be accounted for in terms of the overlap between the two measurements. The Overlap Hypothesis, as a way of accounting for the relationship between measurements, helps in showing that many of the correlations in longitudinal series may be thought of as indices of the amount of growth which has taken place by a particular age or developmental period rather than as predictions of the amount of growth which will take place between the earlier and the later measurement.

Several statistical formulations of the Overlap Hypothesis have been considered and the assumptions underlying these formulations briefly discussed. When certain of these assumptions are met, it is possible to use cross-sectional or normative data to estimate the relationships to be expected in a longitudinal study. Thus the two types of studies (longitudinal and normative) may be used to support each other and both may be used to describe the development of a particular characteristic.

Using normative data on different samples at each age and the longitudinal correlations from each age to maturity, it is possible to describe the development of height from birth to maturity. Both types of data reveal the very rapid early development of height, followed by a period of slow but steady growth, followed by the adolescent spurt in height growth. Both types of data show the characteristic earlier adolescent growth spurt of girls as compared with boys and the influence of the physiological processes at puberty in controlling further height growth.

The influence of environment on height growth is most clearly demonstrated when extreme environments are studied. It has been proposed that the effect of the environment is greatest in the period of most rapid normal development of the characteristic and it is least in the periods of least rapid normal development. The evidence bearing on this point has been considered for both animal and human data.

REFERENCES

Anderson, J. E., 1939. The limitations of infant and preschool tests in the measurement of intelligence. *J. of Psychol.*, **8**, 351–379.
Baldwin, B. T., 1921. The physical growth of children from birth to maturity. *Univ. of Iowa Stud. Child Welfare*, **1**, No. 1.
Boas, F., 1911. Changes in bodily form of descendants of immigrants. Washington: U. S. Senate Document, No. 208.
Burt, C., 1955. The evidence for the concept of intelligence. *Brit. J. of Ed.* **25**, 158–177.
Dearborn, W. F., and Rothney, J. W. M., 1941. Predicting the child's development. Cambridge, Mass.: Sci-art Publishers.
Dreizen, S., Currie, C., et al., 1953. The effects of nutritive failure on the growth patterns of white children in Alabama. *Child Develpm.*, **24**, 189–202.
Dreizen, S., Currie, C., Gilley, E. J., and Spies, T. D., 1950. The effect of milk supplements on the growth of children with nutritive failure: II. Height and weight changes. *Growth*, **14**, 189–211.
Elderton, E. M., and Pearson, K., 1915. The relative strength of nurture and nature. London: Cambridge Univ. Press.
Greulich, W. W., 1941. Some observations on growth development of adolescent children. *J. Pediatrics*, **19**, 302–314.
Hammond, J., 1959. Progress in the physiology of farm animals. Supplement. London: Butterworths Scientific Publications.
Husén, T., 1959. Psychological twin research. Stockholm: Almquist and Wicksell.
Jackson, C. M., 1936. Recovery of rats after a prolonged supression of growth by dietary deficiency of protein. *Amer. J. Anat.* **58**, 179–93.
Jackson, C. M., 1937. Recovery of rats upon refeeding after prolonged suppression of growth by underfeeding. *Anat. Rec.* **68**, 371–81.
Jones, H. E., 1948. Motor performance and growth. Stanford, California: Stanford Univ. Press.
Jones, M. C., and Bayley, N., 1950. Physical maturing among boys as related to behavior. *J. of Ed. Psychol.*, 129–148.
Kelley, T. L., 1924. Statistical method. New York: Macmillan.
Keys, A., Henschel, A., et al., 1950. The biology of human starvation. Minneapolis: Univ. of Minnesota Press, 2 volumes.
Krogman, W. M., 1950. A handbook of the measurement and interpretation of height and weight in the growing child. *Monogr. Soc. Res. Child Develpm.*, **13.**
Martin, W. E., 1953. Basic body measurements of school age children. Washington: U. S. Dept. of Health, Education, and Welfare.
McCay, C. M., Crowell, M. F., and Maynard, L. A., 1935. The effect of retarded growth upon the length of life span and upon the ultimate body size. *J. of Nutrition*, **10**, 63–70.

McCay, C. M., Barnes, L. L., Maynard, L. L., and Sperling, G., 1939. Retarded growth, life span, ultimate body size and age changes in the albino rat after feeding diets restricted in calories. *J. of Nutrition*, **18**, 1–13.

McMeekan, C. P., 1940. Growth and development in the pig, with special reference to carcass quality character: II. The influence of the plane of nutrition on growth and development. *J. of Agric. Sci.*, **30**, 387–436.

McNemar, Q., 1949. Psychological statistics. New York: Wiley.

Newman, H. H., Freeman, F. N., and Holzinger, K. J., 1937. Twins: A study of heredity and environment. Chicago: Univ. of Chicago Press.

Olson, W. C., 1955. Child development. Boston: Heath.

Rusch, Reuben R., 1956. The cyclic pattern of height growth from birth to maturity. Michigan State University, Ph.D. Dissertation.

Sanders, B. K., 1934. Environment and growth. Baltimore: Warwick and York.

Scammon, R. E., 1930. "The measurement of the body in childhood," in Harris, J. A., Jackson, C. M., Patterson, D. G., and Scammon, R. E., The measurement of man. Minneapolis: Univ. of Minnesota Press.

Shuttleworth, F. K., 1939. The physical and mental growth of girls and boys age six to nineteen in relation to age at maximum growth. *Monogr. Soc. Res. Child Develpm.*, **4**, No. 3.

Simmons, K., 1944. The Brush Foundation study of child growth and development. II: Physical growth and development. *Monogr. Soc. Res. Child Develpm.*, **9**, No. 1.

Stolz, H. R., and Stolz, L. M., 1951. Somatic development of adolescent boys. New York: Macmillan.

Tanner, J. M., 1955. Growth at adolescence. Springfield, Illinois: C. C. Thomas.

Tuddenham, R. D., and Snyder, M. M., 1954. Physical growth of California boys and girls from birth to 18 years. *Child Develpm.*, **1**, No. 2. Berkeley: Univ. of California Press, pp. 183–364.

Wilson, E. B., 1935. Heights and weights of 275 public school girls for consecutive ages 7–16 years inclusive. *Proceedings of the National Academy of Sciences* **21**: 633–634.

Wooley, H. B., 1926. An experimental study of children at work and in school between the ages of fourteen and eighteen years. New York: Macmillan.

*Chapter Three*

# INTELLIGENCE

Since the development of the Stanford-Binet Test in 1905 and the Army Alpha Test during World War I, these tests and many other intelligence and aptitude tests have been administered to millions of persons in this country and abroad. Fortunately, in addition to the many normative studies of different age groups, some studies have been done in which these tests have been readministered to the same individuals over periods of time ranging from one year to as long as thirty years. It is the longitudinal data from these studies that enable us to examine the stability of general intelligence as well as specific aptitudes. These studies in combination with other research help us to determine some of the conditions which promote stability as well as change in these tested characteristics.

In this chapter we shall attempt to show the consistency of the results from a number of longitudinal studies and will then analyze a representative study in greater detail. We shall attempt to describe the extent to which intelligence test results achieve stability at selected ages and shall consider some of the environmental conditions which affect both the level of tested intelligence as well as the stability of such measures. Finally, we shall compare the results for general intelligence with the results for selected aptitudes.

GENERAL INTELLIGENCE

Tests of general intelligence such as the Stanford-Binet, Wechsler-Bellevue, Otis, Terman, Kuhlman-Anderson, etc., have been read-

*52*

ministered to groups of children in the major longitudinal studies. The Harvard Growth Study (Anderson, 1939), the University of Chicago Study (Freeman and Flory, 1937), the California Guidance Study (Honzik et al., 1938), the Berkeley Growth Study (Bayley, 1949), the Brush Foundation Study (Ebert and Simmons, 1943), and the Fels Foundation Study (Sontag et al., 1958) have followed groups of children for periods ranging from 5 to 21 years with these and other general intelligence tests.

The results of these different longitudinal studies of intelligence have been plotted to show the relationship between test scores at each age and the test scores at a criterion age. In Chart 1 the criterion age is 10, whereas the criterion age is 16 to 18 in Chart 2. It will be noted that the curves in each chart are very similar in form, although some studies show a lower level of relationship than do others. The curves in Chart 1 are very similar and a single general trend clearly emerges. In Chart 2 a similar general trend also emerges, although the Anderson (1939) study deviates markedly from the other three studies. A major difference between the other studies and the Harvard Study, on which the Anderson correlations are based,

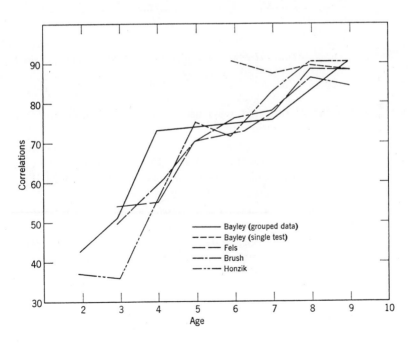

*Chart 1. Correlations between Intelligence at Each Age and Intelligence at Age 10.*

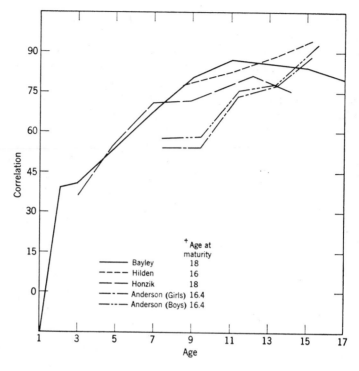

*Chart 2.    Correlations between Intelligence at Each Age and Intelligence at Maturity.*+

was the number of different tests used.    At least 10 different tests were used in the Harvard Study, most of them group tests.    The relationships between these tests, even when given within the same week to a group of students at age 10, is of the order of +.80, which is lower than the reliability of the tests, especially of the Stanford-Binet test which is approximately +.90 at that age.    We attribute the differences between the Harvard Study results and the others to the effect of the varying content of the tests used rather than to any fundamental difference in the samples used or in the underlying pattern involved in the development of intelligence.    We have omitted the Harvard data in Chart 2a.*    However, before leaving the Harvard data, we wish to point out the similarity in the curves for boys and

* It is of interest to note that when the Harvard Study correlations are corrected for the differences between tests by using the average r of .80 as the estimated correlation between the tests, the corrected correlations based on the Spearman Brown attenuation formula approach the value of the r's found in the other studies.

girls.   At least during the ages 7 to 16 the correlations for both sexes are so similar that there is little justification for separate analyses of longitudinal data on intelligence for girls and boys.   The other studies represented in Charts 1 and 2 are for girls and boys combined.

Each of the studies represented in Charts 1 and 2 is based on a particular sample of individuals that participated in a longitudinal study over periods of time ranging from 5 to 21 years.   Such a sample may differ from the general population not only in central tendency but also in variability.   Since the variability of the sample will affect the magnitude of the correlations, we have used Kelley's (1924) formula to correct all the correlations in the two charts to the standard deviations reported by Terman and Merrill (1937) for a large representative sample of children in the United States.   The corrections are made only for the standard deviation at the criterion age since no satisfactory formula is available for correcting both standard deviations.   The corrected correlations are shown in Charts 1a and 2a.

The similarity of the different curves of relationship in Charts 1a and 2a suggests that although these studies were made at different

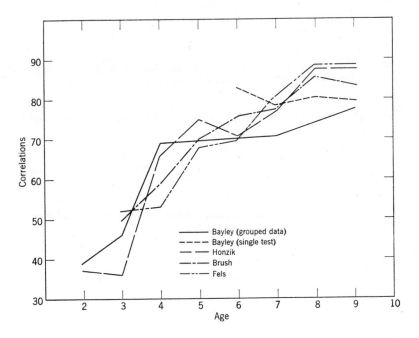

*Chart 1a.   Correlations between Intelligence at Each Age and Intelligence at Age Ten Corrected to a Common Terminal Variability.*

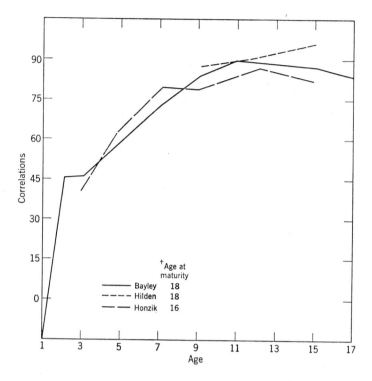

*Chart 2a.    Correlations between Intelligence at Each Age and Intelligence at Maturity+
Corrected to a Common Terminal Variability.*

times and under quite different conditions, the research results begin to delineate a single curve of intelligence development.    This single line approximates very closely the results of different longitudinal studies done with different groups of children, in different parts of the country, with different examiners, and at different times.    The California study was done in the period from the early 1930's to the late 1940's in Oakland, California, the Brush study from the early 1930's to the early 1940's in Cleveland, Ohio, and the Fels study was done from the 1940's to the 1950's in Yellow Springs, Ohio.    The consistency of these data under such different conditions suggests that general intelligence develops in an exceedingly *lawful* way and that the discovery of the underlying nature of this development is worthy of our systematic efforts.

It is evident that we can accept this curve as a good approximation of the interrelationships among similar intelligence tests given at various ages from 1 to 18 years of age.    Although the values may fluctuate somewhat from sample to sample, much of the variation may be ascribed to the difference in variability of particular samples as contrasted with the general population.

The curves we have drawn in Chart 2a show a period of rapidly increasing correlations to about age 9, then a period of slowly increasing correlations from ages 9 to 16. The next section will attempt to explain this characteristic curve.

SELECTION OF ONE STUDY FOR FURTHER ANALYSIS

In order to analyze the development of general intelligence in greater detail we selected Bayley's (1949) study. This represents a most careful and precise longitudinal study of a group of children from birth to age 18. Bayley has a well-defined sample. She reports her data in great detail and uses only five different intelligence tests throughout the study.

She used the California First Year Tests until 15 months, the California Preschool Tests until age 5, then used the different forms of the Stanford-Binet from ages 6 to 12 and at 14 and 17, and the Wechsler-Bellevue at ages 16 and 18. She used the Terman-McNemar Test at ages 13 and 15. Fortunately, she made greatest use of the Stanford-Binet Tests, which were administered under excellent conditions by well-trained testers.

Bayley's data approximate rather closely the various curves in Chart 2a and it is likely that generalizations derived from these data are generally applicable. Because we wished to reduce the complications introduced by using different tests, we have used the Stanford-Binet test at age 17 as the criterion measure in most of our analyses and have omitted the Wechsler-Bellevue Test at ages 16 and 18 and the Terman-McNemar Test at ages 13 and 15. However, in the grouped data analyses we have made use of the complete set of results reported by Bayley. We have plotted various analyses of this study in Chart 3.

Intelligence measured at age 1 has a zero correlation with intelligence measured at age 17. Intelligence measured at age 2 has a correlation of $+.41$ with the intelligence measured at age 17. By age 4, the correlation with the measurement at age 17 has increased to $+.71$ and by age 11 it has increased to $+.92$. Thus, we see the characteristic rapid increase in the correlations between the criterion measure and measurements made in the early years and a less rapid increase in the relationships after age 4. From these results we would conclude (with

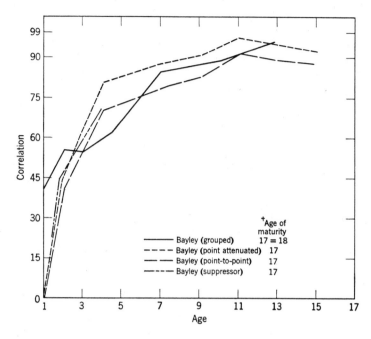

*Chart 3.  Correlations between Intelligence at Each Age and Intelligence at Maturity*[†]
*Based on Several Treatments of the Longitudinal Data (Adapted from Bayley,
1949).*

Bayley) that the intelligence measured at age 1 by the California
First Year Test is not predictive of the kinds of intelligence being
measured at age 17 by the Stanford-Binet.

In the foregoing we have been dealing with point to point relationships, that is, the correlations between a test score at one time and a
test score at another time.   Another way of looking at these data is to
determine the relationship between the evidence on intelligence test
performances up to a particular point in time and the evidence at a
later point in time.   For example, if we have evidence from intelligence
tests at ages 6 months, 1 year, $1\frac{1}{2}$ years, 2 years, and $2\frac{1}{2}$ years, how
well will some combination of these measurements predict intelligence
test scores at age 17?

This question has, in part, already been studied by Bayley (1949),
who converted the I.Q. scores into standard scores, summarized the
results from three testings (two at ages 17, 18) and determined the
intercorrelations among these combined or grouped test results.
Chart 3 compares the correlations for the grouped test results with

those obtained by using point-to-point data. It may be seen that correlations for the grouped data are considerably higher than the correlations obtained from the point-to-point data at most points except ages 3 to 6 where the point-to-point correlations are higher and ages 10 to 11 where both are approximately the same. It is especially significant to note that intelligence as measured by the California First Year tests at ages 10, 11, and 12 months correlates +.41 with intelligence measured at ages 17 and 18 years (in contrast to the zero correlation between test scores at 13 months and test scores at age 17). The correlation between the scores on tests administered at 18, 21, and 24 months and at 17 and 18 years is now +.55 instead of +.41. It is especially in the early years that the combining of the scores from several administrations of the test has a marked effect on the correlations with intelligence measured at maturity (17 + 18). We are of the opinion that the main reason why combining scores from several administrations of the test produces higher correlations is because it corrects for the unreliability of the test, which is especially low in the early years where it is difficult to secure ideal cooperation from the examinee.

In order to check this point we have corrected the Bayley point-to-point correlations for attenuation by using Anderson's (1939) and Terman and Merrills' (1937) estimates of the reliability of the Stanford-Binet Test at different ages. In Chart 3, it will be noted that with the exceptions of ages 1 and 4 the point-to-point correlations corrected for attenuation are very similar to the correlations based on the grouped data. Thus, combining the results from several test administrations reduces the error variation in the test scores and yields essentially the same result as would be obtained by the use of a highly reliable test.*

Another approach to the analysis of these data begins with the recognition that the nature of the qualities measured by existing intelligence tests changes from the infant tests to the tests used at maturity. It is clear that the tests in the first 18 months are highly saturated with motor and physical developmental skills. This has been pointed up by Maurer (1946), Hofstaetter (1954), and Cronbach (1960, p. 166). The tests used at ages 17 and 18 are highly saturated with cognitive skills and verbal ability, whereas the tests used before age 2 are largely psychomotor. The tests used at ages 2 and 3 are likely to be combinations of the two types of abilities.

---

* It would follow from these data that especially in the first few years no decisions of any significance should be made about a child on the basis of a *single* measurement of intelligence.

If the types of abilities tested at ages 18 months or earlier and those tested at age 10 and later are so different, we would expect the correlation between psychomotor tests (before 2 years) and the cognitive and verbal ability tests (after age 10) to be very near zero. This is exactly what has been found in a number of studies (Cavanaugh et al., 1957; Cattell, 1931; and Anderson, 1939).

There is little reason to expect scores on psychomotor tests to predict scores on tests of cognitive and verbal ability since even when psychomotor tests and verbal ability tests are given at about the same time the correlation between them approaches zero (Cronbach, 1960, p. 274).

The tests given at ages 2 and 3 are composite tests which include both psychomotor as well as cognitive-verbal tasks. Since these tests contain both types of tasks, we may expect them to be moderately correlated with tests before age 2 as well as with tests after age 10. It will be found in Table 3.1 that the tests at 2 and 3 years are as highly related with tests after 10 years as they are with tests before 2 years.

The implication that follows from Table 3.1 is that two kinds of abilities are being tested at ages 2 and 3. One is a psychomotor type of ability which is highly related to the tests used at ages 1 to $1\frac{1}{2}$ years. Other abilities, which we have termed cognitive, are also included in the tests used at ages 2 and 3 and these are related to abilities tested after age 10 but not to the abilities tested before age 2. Thus the tests at ages 2 and 3 are complex tests measuring two very distinct types of development. If these tests could be purified so as to contain only psychomotor tests, they should correlate more highly with the infant tests given before age 2. On the other hand, if they could be purified so as to contain only cognitive tasks, they should correlate

*Table 3.1. Longitudinal Relationships among Tests Classified as Psychomotor, Psychomotor-Cognitive, and Cognitive Verbal*

(*Adapted from Bayley*, 1949)

| | Psychomotor-Cognitive | | Cognitive-Verbal | |
|---|---|---|---|---|
| *Psychomotor* | *2 Years* | *3 Years* | *11 Years* | *17 Years* |
| 1 year† | +.47 | +.41 | +.02 | +.002 |
| 1½ years | +.50 | +.54 | +.11 | +.20 |
| *Psychomotor-Cognitive* | | | | |
| 2 years | | | +.43 | +.41 |
| 3 years | | | +.48 | +.56 |

† Thirteen months.

*Table 3.1a. Multiple Correlations to Determine the Effect of Suppressing Psychomotor Elements in Psychomotor-Cognitive Tests*

(*Adapted from Bayley, 1949*)

| Multiple Correlations | Simple Correlations |
|---|---|
| *R* 17 years:: 2 years: 1 year† = +.47 | *r* 17 years–2 years = +.41 |
| *R* 17 years:: 3 years: 1 year† = +.61 | *r* 17 years–3 years = +.56 |
| *R* 17 years:: 2 years: 1½ years = +.41 | *r* 17 years–2 years = +.41 |
| *R* 17 years:: 3 years: 1½ years = +.62 | *r* 17 years–3 years = +.56 |

† Thirteen months.

relatively highly with the tests given at age 17.   Maurer (1946) made a study of the relation between preschool test items and later measures of intelligence.   She found two types of items.   One set of preschool test items correlated much more highly with intelligence measures at age 15 than did the other set.

Although we do not have the data necessary for the purification of these tests by item analysis procedures, we are, however, in a position to determine the effect of suppressing the psychomotor or the cognitive portions of tests given at ages 2 and 3 by means of a statistical formula suggested by Guilford and Michael (1948) and McNemar (1949). The use of this multiple correlation formula to suppress the effect of an unwanted source of variation has been discussed at some length by Guilford and Michael (1948).

It will be noted in Table 3.1a that essentially what had been predicted did take place.   The correlations between the tests at age 17 and those at ages 2 and 3 are increased by holding constant or suppressing the results of the test at age 1.   When the test at 1½ years is held constant, the effect on the correlations is not as clear.

When we suppress the psychomotor portion of the tests at ages 2 and 3 and correlate the resulting measures with the cognitive-verbal tests at age 17, the correlations are slightly higher than the corresponding point-to-point correlations and are very close to those obtained by correcting the point-to-point correlations for attenuation (see Chart 3).

From Chart 3 we may conclude that the correlation between intelligence (when ideally measured) at age 3 and age 17 is about +.65, between intelligence measured at age 5 and age 17 is about +.80, and between intelligence measured at age 8 and age 17 is almost +.90. After age 8, the correlations between repeated tests of general intelligence should be between +.90 and unity.

In the analysis of height measurements, we found that the correlations between height measurements at each age and height at maturity (age 18) was a close fit with the percent of the mature height reached by each age.   This suggested that Anderson's Overlap Hypothesis might serve as the explanation for the correlation between any two age measurements especially since the correlations between initial measures of height and *gains* to maturity were very low or approached zero.

Before attempting to apply the Overlap Hypothesis to intelligence test data, we may ask whether the relationship between intelligence at each age and the *changes* in intelligence to age 17 are also low.   In Chart 4, we have plotted the correlations between I.Q. at each age and the changes in I.Q. from that age to age 17.   The relationships between initial scores and changes to age 17 are relatively low until age 7. They approach zero at ages 4 to 7, whereas at the earlier periods they

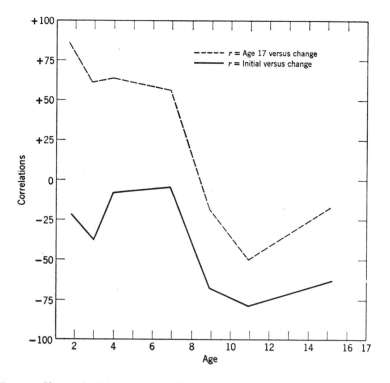

Chart 4.   *Changes in Intelligence from Each Age to Age* 17 *Correlated With Intelligence at Age* 17 *and with Intelligence at the Beginning of the Change Period (Bayley, 1949).*

are less than −.40. We are not clear as to why the correlations between initial scores and changes tend to be negative other than for a general ceiling effect in which persons with initially high scores tend to make smaller changes than persons with initially low scores. In any case until age 7 the correlations between initial scores and changes are relatively low. They are significantly different from zero only after age 7.*   Much the same patterns of low or zero correlations between initial position and changes in I.Q. were reported by Roff (1941, p. 385) who states

. . . the correlations between test scores at one age and gain in scores at a later age are as likely to be negative as positive, and fluctuate around zero . . . . These results indicate that the so-called "constancy of the I.Q." is due primarily to the retention by each child of the skills and knowledge which determined his scores in earlier years, and is not due at all to correlation between earlier scores and later gains or increments.

Height measurements are expressed in absolute and equal units (inches or centimeters) and this may explain why we obtained such a close fit between the height correlations and the estimates obtained from the Overlap Hypothesis.   Since I.Q. and Mental Age are relative units which are not necessarily equal, we have searched for units of development which will more nearly approximate the type of absolute and equal unit we find in physical measurements.   We have tried the Overlap Hypothesis using chronological age and mental age means as units.   Both of these seriously underestimate the correlations reported in the Bayley Study (1949) (see Chart 5).   It should be noted that the Overlap Hypothesis using chronological age or mental age does fit the correlations during the period 2 through 10† even though these ratios do not fit the data for the longer period 2 though 17.

* In contrast with the low correlations between initial scores and changes are the relatively high correlations between final scores (age 17) and the changes.   These serve to make clear that the low correlations between initial score and changes are not to be entirely explained in terms of the low reliability of change scores.
† The Bayley grouped data in Chart 1a fit rather closely the ratio of chronological ages; thus when using age 10 as criterion the observed and expected correlations are as follows:

| Ages | Observed r | Expected r |
| --- | --- | --- |
| Age 2–10 | .42 | .45 |
| Age 4–10 | .73 | .63 |
| Age 6–10 | .74 | .77 |
| Age 8–10 | .82 | .84 |

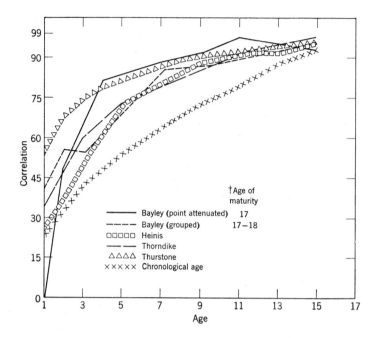

*Chart 5. Correlations between Intelligence at Each Age and Intelligence at Maturity†
Contrasted With Selected Growth Curves for Intelligence.*

One possible explanation for the close fit between chronological age units and the correlations for intelligence in the 2 to 10 age period is the fact that this is the period when the most rapid changes take place in general intelligence and the correlational pattern most nearly approaches a straight line. However, in the period ages 2 to 17 the line is more nearly a parabolic function tending to a plateau after age 9.

When, however, we turn to various attempts to represent the development of intelligence in absolute units, these usually show a very rapid development in the early years from 2 to about 7, with a relatively small amount of development taking place after age 9. Using the developmental curves proposed by Thurstone (1928), Thorndike (1927), and Heinis (1924) in which absolute units are approximated, we find that the Overlap Hypothesis using these units fits very closely to the results obtained by Bayley using either the grouped data or the point-to-point correlations corrected for attenuation (see Chart 5). These developmental curves based on absolute units also have a close fit with the curves representing the different longitudinal studies.

In Chart 5 it may be seen that the three methods of estimating the absolute development of intelligence are very similar after age 7 although they give slightly different results during the period age 2 to age 7. The Thorndike curve most nearly approximates Bayley's data although it somewhat underestimates the correlations in the first two years. We conclude that the absolute scales for the development of intelligence when related to Anderson's Overlap Hypothesis does account for the increasing correlations between intelligence test scores as the measurements approach a terminal or criterion age.

Before leaving this point we may ask whether the general trends in the development of intelligence shown in Chart 5 are an accurate picture of the many longitudinal studies in which two or more measurements based on the same intelligence test were used. We have not included studies which used different tests in the follow-up or studies based on samples with variable ages. In Table 3.2 we have summarized 13 studies by showing the correlations between the measurements made at different ages. In Column B, we have shown the correlations to be expected on the basis of the Thorndike curve of the development of intelligence. Since this curve was drawn with the assumption of perfect reliability in the measurements, we have estimated the correlation to be expected if the reliability of the measurements is that reported for the specific tests used (column C). In the studies, where the sample variability is different from the variability of the normative population, we have corrected the estimated correlation for the curtailed variability of the sample (column D). For the 13 observed correlations shown in Table 3.2, nine do not differ significantly from the estimated values based on Thorndike's curve.

The samples on which Table 3.2 is based are generally in the "normal" range of intelligence. In order to determine whether our estimates also fit the results of longitudinal studies based on superior and below-normal samples, we have summarized a few of these studies in Table 3.3. Here it will be seen that the estimated correlations (column B) are in all cases higher than the observed correlations (column A). However, when the estimated correlations are reduced to account for both test unreliability and the curtailed variability of these extreme samples, the observed correlations (column A) and the estimated correlations (column D) are more nearly similar. Out of the five comparisons, only one is significantly lower than the theoretically expected value.

We would conclude that the Overlap Hypothesis, when related to an absolute scale of intelligence such as Thorndike's, does account for the correlations generally found in longitudinal studies of intelligence. We are of the opinion that further work on the theory and methods of

Table 3.2. Test-Retest Studies of Intelligence of Normal Children

| Author and Date | Sample | Test | N | Ages | Observed | Theoretically Expected | | |
| | | | | | A | B When Test Reliability Perfect | C Reduced by Actual Test Reliability | D Reduced to Account for Both Reliability and Variability |
|---|---|---|---|---|---|---|---|---|
| Bradway (1944) | School children | Stanford-Binet ('37) | 52 | 2 and 3 versus 12 and 13 | .58 | .59 | .54 | .59 |
| Skodak and Skeels (1949) | Adopted children† | Stanford-Binet ('16) | 100 | 4½ versus 13½ | .59 | .72 | .66 | .62 |
| Freeman, Holzinger and Mitchell (1928) | School children | Stanford-Binet ('16) | 74 | 8 versus 12 | .68 | .91 | .82 | .78 |
| Skodak and Skeels (1945) | Adopted children† | Stanford-Binet ('16) | 139 | 4½ versu 7½ | .72 | .83 | .76 | .70 |
| Hirsch (1930) | School children | Otis | 160 | 7 and 8 versus 12 and 13 | .80 | .87 | .81 | |
| Goldin and Rothschild (1942) | School children | Henmon-Nelson | 54 | 10 versus 14 | .83 | .93 | .84 | .84 |
| Winons (1949) | School children | Cal. Test of Ment. Maturity | 169 | 13½ versus 17 | .68* | .97 | .87 | .79 |
| Layton (1954) | High School students | Ace Psych. Exam. High Sch. Ed. | 2169 | 14 versus 17 | .80** | .98 | .88 | .88 |
| Townsend (1944) | School children | Kuhlman-Anderson | 59 | 6 versus 9 | .65* | .88 | .79 | |
| Knezevich (1946) | High school students | Henmon-Nelson | 113 | 15 versus 17 | .70 | .98 | .88 | .76 |
| Pintner and Stanton (1937) | School children | Thorndike CAVD | 59 | 8 versus 10 | .76 | .94 | .85 | |
| Wentworth (1926) | School children | Dearborn | 575 | 6 versus 7 | .72* | .95 | .81 | .77 |
| Stalnaker and Stalnaker (1946) | High school students | SAT Verbal Test | 2000 | 16 versus 17 | .94 | .99 | .94 | |

† Adopted before 6 months.
* Significant at .05 level.
** Significant at .01 level.

Table 3.3. Test-Retest Studies of Intelligence of Exceptional Children

| | | | | | Correlations | | | |
|---|---|---|---|---|---|---|---|---|
| | | | | | Observed | Theoretically Expected | | |
| | | | | | A | B | C | D |
| | | | | | | When Test Reliability Is Perfect | Reduced by Actual Test Reliability | Reduced to Account for Both Reliability and Variability |
| Author and Date | Sample | Test | N | Ages | | | | |
| **Superior** | | | | | | | | |
| Burks (1930) | Gifted children | Stanford-Binet ('16) | 54 | 10 versus 16 | .81 | .92 | .84 | .73 |
| Katz (1942) | Preschool gifted children | Stanford-Binet ('37) | 268 | 3 versus 5 | .62 | .82 | .75 | .62 |
| Traxler (1934) | Lab. school children | Otis | 85 | 13½ versus 16½ | .68* | .97 | .90 | .82 |
| **Retarded** | | | | | | | | |
| Kirk (1958) | Mentally retarded children living at home | Stanford-Binet ('37) | 26 | 4 versus 7 | .81 | .82 | .73 | .73 |
| Kirk (1958) | Mentally retarded children in an institution | Stanford-Binet ('37) | 12 | 4½ versus 7 | .51 | .86 | .77 | .57 |

* Significant at .05 level.

absolute scaling may yield even closer approximations to observed data not only for general intelligence but for many other human characteristics.

Using the results in Chart 5, we may now begin to describe the development of general intelligence. Using either Bayley's correlation data ($r^2$) or the Thorndike absolute scale (both of which yield essentially the same results), it is possible to say, that in terms of intelligence measured at age 17, at least 20% is developed by age 1, 50% by about age 4, 80% by about age 8 and 92% by age 13. Put in terms of intelligence measured at age 17, from conception to age 4, the individual develops 50% of his mature intelligence, from ages 4 to 8 he develops another 30%, and from ages 8 to 17 the remaining 20%. This differentially accelerated growth is very similar to the phenomenon we have noted in Chapter 2 with regard to height growth.

With this in mind, we would question the notion of an absolutely constant I.Q. Intelligence is a developmental concept, just as is height, weight, or strength. There is increased stability in intelligence measurements with time. However, we should be quick to point out that by about age 4, 50% of the variation in intelligence at age 17 is accounted for. This would suggest the very rapid growth of intelligence in the early years and the possible great influence of the early environment on this development.

We would expect the variations in the environments to have relatively little effect on the I.Q. after age 8, but we would expect such variation to have marked effect on the I.Q. before that age, with the greatest effect likely to take place between the ages of about 1 to 5.

## THE EFFECT OF VARIATIONS IN ENVIRONMENT ON INTELLIGENCE

We may assume that an individual is born with a nervous system and physiological makeup which are the bases on which general intelligence is developed. Individuals may vary considerably in these characteristics at birth, and this variation is undoubtedly a significant factor in determining the potential of the individual for the development of general intelligence.

Perhaps our clearest evidence for the inheritance of general intelligence comes from the various studies of twins reared together and reared apart. In Table 3.4 we have summarized the various studies on the intelligence measurements of twins as well as siblings. Here it will be noted that the identical twins, when reared apart, correlate

*Table 3.4. Correlational Studies of the Intelligence of Twins and Siblings Reared Together and Reared Apart*

|  | Group Tests of Intelligence | | | Individual Tests of Intelligence | |
| --- | --- | --- | --- | --- | --- |
|  | Burt (1958) | Newman, Freeman, Holzinger (1937) | Husén (1959) | Burt (1958) | Newman, Freeman, Holzinger (1937) |
| **Identical Twins** | | | | | |
| Reared together | .94 | .92 | .90 | .92 | .91 |
| Reared apart | .77 | .73 | | .84 | .67 |
| **Nonidentical Twins** | | | | | |
| Reared together | .54 | .62 | .70 | .53 | .64 |
| **Siblings** | | | | | |
| Reared together | .52 | | | .49 | |
| Reared apart | .44 | | | .46 | |
| **Unrelated Children** | | | | | |
| Reared together | .28 | | | .25 | |

+.67 to +.84 as compared with +.90 to +.94 for identical twins reared together. Similarly, the siblings reared together correlate slightly higher than do the siblings reared apart. Here then, we do have evidence that similar hereditary makeup when accompanied by similar environments (at least home environments) does result in very similar levels of general intelligence, whereas similar hereditary makeup accompanied by dissimilar environments results in somewhat different levels of general intelligence.

This point is further clarified by Anastasi (1958, p. 299) in a table showing the differences in I.Q.'s of identical twins reared apart as related to environmental differences (See Table 3.4a). It is especially noteworthy that the differences in I.Q. for identical twins separated during the first three years are highly related to the differences in educational advantage (+.79) but have only moderate relationships with the difference in social and physical advantages in the environments of the separated twins (+.51, +.30).* Thus, if the identical twins are separated but placed in very similar environments, it is likely that they will have very similar intelligence tests scores, whereas if

* Reported by Newman, Freeman, Holzinger (1937, p. 340).

*Table 3.4a. Environmental Differences and I.Q. Differences for Identical Twins Reared Apart*

*(Anastasia, 1958)*

Environmental Differences

| Case No. | Sex | Age at Separation | Age at Testing | 1. Years of Schooling | 2. Educational Advantages | 3. Social Advantages | 4. Physical Advantages | I.Q. Differences† |
|---|---|---|---|---|---|---|---|---|
| 11 | F | 18 mo. | 35 | 14 | 37 | 25 | 22 | 24 |
| 18 | M | 1 yr. | 27 | 4 | 28 | 31 | 11 | 19 |
| 4 | F | 5 mo. | 29 | 4 | 22 | 15 | 23 | 17 |
| 8 | F | 3 mo. | 15 | 1 | 14 | 32 | 13 | 15 |
| 2 | F | 18 mo. | 27 | 10 | 32 | 14 | 9 | 12 |
| 1 | F | 18 mo. | 19 | 1 | 15 | 27 | 19 | 12 |
| 17 | M | 2 yr. | 14 | 0 | 15 | 15 | 15 | 10 |
| 12 | F | 18 mo. | 29 | 5 | 19 | 13 | 36 | 7 |
| 6 | F | 3 yr. | 59 | 0 | 7 | 10 | 22 | 8 |
| 9 | M | 1 mo. | 19 | 0 | 7 | 14 | 10 | 6 |
| 10 | F | 1 yr. | 12 | 1 | 10 | 15 | 16 | 5 |
| 5 | F | 14 mo. | 38 | 1 | 11 | 26 | 23 | 4 |
| 16 | F | 2 yr. | 11 | 0 | 8 | 12 | 14 | 2 |
| 13 | M | 1 mo. | 19 | 0 | 11 | 13 | 9 | 1 |
| 15 | M | 1 yr. | 26 | 2 | 9 | 7 | 8 | 1 |
| 7 | M | 1 mo. | 13 | 0 | 9 | 27 | 9 | −1 |
| 14 | F | 6 mo. | 39 | 0 | 12 | 15 | 9 | −1 |
| 3 | M | 2 mo. | 23 | 1 | 12 | 15 | 12 | −2 |
| 20 | F | 1 mo. | 19 | 0 | 2 | ? | ? | −3 |
| 19 | F | 6 yr. | 41 | 0 | 9 | 14 | 22 | −9 |

† Difference in favor of the individual with the educational advantage is shown as positive.

placed in very different environments, their intelligence test scores will be quite different.*

Various workers have attempted to determine the proportions of the variance attributable to heredity and environment. Woodworth

* We have divided the separated twins into two groups. For one group of 11 pairs, each pair of separate twins had very similar educational environments. The rank correlation for their I.Q. scores was +.91, whereas for the eight pairs that had the least similar educational environments, the rank correlation for their I.Q. scores was only +.24.

(1941) estimates 60% attributable to heredity, Newman, Freeman, and Holzinger (1937) estimate 65% to 80% attributable to heredity, Burks (1928) estimates 66% to heredity, Leahy (1935) estimates 78%, whereas Burt (1958) estimates 77% to 88% attributable to heredity. We do not propose to attempt to settle this controversy other than to recognize that although the estimates vary, all are apparently agreed that some portion of the variance must be attributed to the effect of the environment in which the children are reared.

We shall consider the evidence as to what constitutes a favorable or abundant environment for the development of intelligence and shall do the same for an unfavorable or deprived environment. We take the view that intelligence is a developmental characteristic in that the mental age or I.Q. compares the general learning of an individual with the progress in the learning of selected samples of behavior made by representative samples of individuals at different ages. It would seem that with such an operational concept of intelligence, the environment could clearly block and retard certain developments in an individual, whereas it is likely (but less clear) that the environment could facilitate and accelerate these developments (Hunt, 1961). If general intelligence is a developmental characteristic and is related to the time it takes the individual to learn various concepts, skills, etc., it would seem reasonable that lack of such learning in one time period may be difficult or impossible to make up fully in another period, whereas unusually excellent learning in one time period is not likely to be lost in a subsequent period.

What then is likely to happen if the individual lives in a deprived or abundant environment (as it affects intelligence) for varying periods of time in his development? Let us assume that the long-term effect of extreme environments may affect the I.Q. to the extent of 1.25 standard deviations on the I.Q. norms, that is, about 20 I.Q. points. This seems a reasonable amount since the three pairs of identical twins reared under the most different environments (see Table 3.4a) had an average difference of 20 I.Q. points, while Sontag (1958) found the individuals in his study changing as much as 20 I.Q. points under what he considered to be favorable and unfavorable environments. This is about the figure cited by Burks (1928) as the effect of extreme environments. We are not firmly convinced of the magnitude of the differences produced by abundant and deprived environments, but do regard 20 I.Q. points as a fair estimate of this amount. With this figure as a first approximation, and with our estimates of the extent of development of intelligence at various ages as determined by the Overlap Hypothesis related to the work of Thorndike (1927) on an

*Table 3.5. Hypothetical Effects of Different Environments on the Development of Intelligence in Three Selected Age Periods*

| | | Variation from Normal Growth in I.Q. Units | | | |
|---|---|---|---|---|---|
| Age Period | Percent of Mature Intelligence | Deprived | Normal | Abundant | Abundant-Deprived |
| Birth–4 | 50 | −5 | 0 | +5 | 10 |
| 4–8 | 30 | −3 | 0 | +3 | 6 |
| 8–17 | 20 | −2 | 0 | +2 | 4 |
| Total | 100 | −10 | 0 | +10 | 20 |

absolute scale of intelligence, we are in a position to set up a hypothetical table of the possible effects of various environments on the development of intelligence.

In Table 3.5 we have hypothesized the possible effects on the individual's intelligence of living under different environments. It will be noted that we are here concerned only with the first 17 years of life. Furthermore, we have assumed that the loss of development in one period cannot be fully recovered in another period. We shall discuss this point later. What we have hypothesized is that extreme environments can have far greater effects in the early years of development than they can in later years. That is, deprivation in the first four years of life can have far greater consequences than deprivation in the ten years from age 8 through age 17. Put in other terms, extreme environments each year in the first four may affect the development of intelligence by about an average of 2.5 I.Q. points per year, whereas extreme environments during the period of age 8 to 17 may have an average effect of only 0.4 I.Q. points per year.

Is there any evidence in support of the values hypothesized in Table 3.5? We will first consider a number of the studies on the effects of deprivation in relation to our hypothesized values, then the studies of the effects of abundance, and finally the studies of individuals who have moved from one type of environment to another. In each study, we cannot be certain of how extreme the environment was, but we will attempt to characterize briefly the environment and show the reported effect in contrast with our hypothesized values. For convenience, we have put the research findings and the hypothesized values in Table 3.6.

In all but one of the studies reported in Table 3.6 the observed differ-

Table 3.6. Effects of Deprivation and Abundance on I.Q. in Contrast with Hypothesized Effects

| Author and Date | Environment | Test | Ages | I.Q. at Initial and Retest | Observed Differences | Hypothesized Change |
|---|---|---|---|---|---|---|
| **Deprivation** | | | | | | |
| Dennis and Najarian (1957) | Orphanage with minimum of adult-child contact | Goodenough, Draw a Man | 4–6 | 93–89 | −4 | −2.5 |
| Newman, Freeman, Holzinger (1937) | Identical twins reared apart; 4 pairs with most extreme environments | Stanford-Binet | Tested at maturity | | 18 | 20 |
| Kirk (1958) | Retarded children in institutions | Stanford-Binet | 4.8–7.3 | 57–51 | −6 | −2 |
| Wheeler (1932) | Isolated mountain children (1930) | | 6 versus 16 | 94.7–73.5 | −21.2 | −3.3 |
| Wheeler (1942) | Less isolated mountain children | | 6 versus 16 | 102.6–81.3 | −21.3 | −3.3 |
| Sontag et al. (1958) | Children with low need achievement | Stanford-Binet | 3–12 | 128–115 | −13 | −5.1 |
| **Abundance** | | | | | | |
| Sontag et al. (1958) | Children with high need achievement | Stanford-Binet | 3–12 | 118–138 | +20 | +5.1 |
| **Deprivation to Improved Environment** | | | | | | |
| Kirk (1958) | Retarded children in institution presented with preschool stimulation | Stanford-Binet | 4.4–7.3 | 61–71.2 | +10.2 | +2.25 |
| Lee (1951) | Negro children born in South and moved to Philadelphia at various ages | | Deprived 1–6 Improved 7–15 | 86.5–92.8 | +6.3 | +2.3 |
| | | | Deprived 1–9 Improved 10–15 | 86.3–89.4 | +3.1 | +1.1 |
| | | | Deprived 1–11 Improved 12–15 | 88.2–90.2 | +2.0 | +0.4 |

ence is greater than our hypothesized changes. The one study in which the observed and hypothesized values are most similar is Newman, Freeman, Holzinger (1937) in which the identical twins with the most radically different educational environments are compared. In this study most of the children had been separated in the first year of life and they were tested at maturity; thus the entire period of growth was considered. In the other studies we have seriously underestimated the amount of change. However, it is of interest to note that the rank order correlation between the observed and hypothesized change is +.95, suggesting that the basic difficulty is not in the general conception of decreasing change with increasing age, but it is in the fixing of the amount of change likely to take place at each age period. It is our hope that future research will establish more accurate estimates of the amount of change which can take place under various environmental conditions at different age periods.

The two studies which we believe to be most crucial in establishing the pattern of change in relation to the environment are those by Kirk (1958) and Lee (1951). In each of these studies, children in contrasting environments were repeatedly tested. In the Kirk study, mentally retarded children in an institution were given a one year preschool experience intended to stimulate their learning. The children were tested prior to the preschool experience at about age 4½. They were retested at the end of the preschool experience, and then again several years later. Another group of children in the institution was used for purposes of contrast. The pre- and post-test scores are shown in Chart 6 for the experimental and contrast groups. It will be seen that with only two exceptions, individuals in the experimental group gained in a rather consistent pattern. The two children in the experimental group who showed decreases in I.Q. consisted of a child with brain damage and a child who appeared to be emotionally disturbed. The children in the contrast group generally decreased in measured intelligence with only two children gaining.*

The Lee (1951) study followed several groups of Negro children with repeated tests until grade 9. In Chart 7, it will be seen that the children who were born in Philadelphia maintained about the same mean scores from grades 1 to 9. The children who were born in the South and moved to Philadelphia by age 6 gained an average of 6½ I.Q. points from grades 1 to 9. The children who were born in the South and moved to Philadelphia by grade 4 gained about 3 I.Q. points from grades 4 to 9, whereas the children who were born in the South

---

* Six of the 15 children in the experimental group showed enough improvement to be released from the institution. None of the children in the contrast group was released.

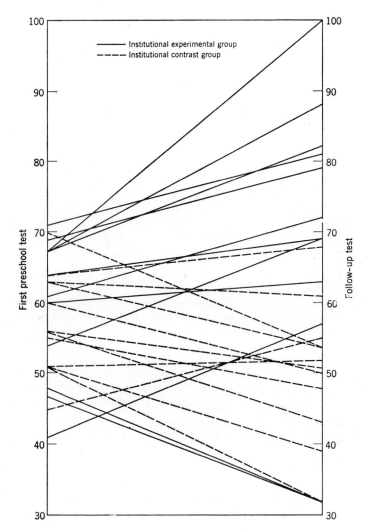

*Chart 6.  Changes in Intelligence for two Groups of Children in an Institution (Adapted from Kirk, 1958).*

and moved to Philadelphia by grade 6 gained only 2 I.Q. points during the period grades 6 to 9.*

* Lee used grade placement of the students in the Philadelphia schools rather than age at entrance.  It is likely that if age samples were used (and these were typical of Southern rural communities), the group would show decreasing levels of intelligence with increase in length of time spent in the South.

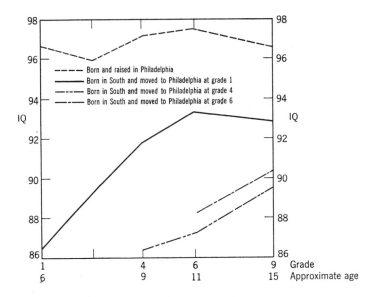

*Chart 7.    Changes in Intelligence for Negro Students Born and Raised in Phildelphia and Negro Students Born in the South and Moving to Philadelphia at Various Grades (Adapted from Lee, 1951).*

The point of the Lee study is the decreasing effect of an improved environment with increasing age.   It is also of interest to note that the greatest changes take place in the first few years in the new environment.   Although there are strong indications that the environment from which the children came in the South could be characterized as a deprived environment for the development of intelligence, we do not believe that the Philadelphia environment should be regarded as an abundant or stimulating one, even though it did represent an improvement over the earlier environment for these children.

### CHARACTERIZATION OF ENVIRONMENTS

We cannot describe in detail precisely what constitutes abundant and deprived environments for the development of intelligence.   We can, however, begin to indicate some of the characteristics of extreme environments on the basis of research studies which demonstrate intelligence test differences in relation to environmental variables and on the basis of inferences derived from the nature of the problems used in intelligence tests.

Since verbal ability represents a very important part of most general intelligence tests, it is likely that environments which include good models of language usage and which encourage the development of language will stimulate the development of general intelligence, whereas environments in which the models of language usage are poor and which discourage language development will retard or block the development of general intelligence (Bernstein, 1960; Milner, 1951; Burks, 1928).

General knowledge about the world around us is also measured by general intelligence tests. To some extent this is included in verbal ability, but in addition there are many intelligence test items which require making distinctions and comparing objects, ideas, etc. Quite obviously, abundant and deprived environments will differ in the opportunities for direct contact and interaction with the world around us and with the vicarious experiences represented by books, pictures, films, television, etc. The findings of intelligence test differences in urban-rural communities (McNemar, 1942; Klinberg, 1931; and Rosca, 1939), the test scores made by children living in isolated villages (Wheeler, 1932, 1942; Sherman and Key, 1932), and the middle-lower class test differences (Eels et al., 1951; McNemar, 1942; Havighurst and Janke, 1944) are probably due in part to this.

Logical reasoning and problem solving are also stressed in most general intelligence tests. Opportunities to solve problems, encouragement to think clearly about a variety of issues, and encouragement to attack problems in and out of school probably also differ in abundant and deprived environments. An environment which restricts these opportunities and which even discourages the individual from attempting to attack and solve problems on his own is likely to retard intelligence development, whereas an environment which encourages problem solving and clear thinking is likely to facilitate the development of intelligence. Some of the differences between the children of different occupational groups (Terman and Merrill, 1937; McNemar, 1942) and the differences between children of parents with varying levels of education (for example, parents who are college graduates versus parents who have completed less than twelve grades) (Bayley, 1954 and Honzik, 1957) are likely to be in part attributable to the opportunities parents give to their children for problem solving and the encouragement and reinforcement they give for clear and logical reasoning.

The nature of the interaction between adults and children is also important in the differentiation between abundant and deprived environments. Obviously, a minimal interaction between adults and children would give little opportunity for the development of the skills and abilities cited here.   This is particularly true in the study by Dennis and Najarian (1957) but it is also in part true of the differences between the intelligence measurements of children in highly authoritarian orphanages and children in their own homes or in foster homes (Spitz, 1945; Levy, 1947; Skodak and Skeels, 1949).   The research of Sontag et al. (1958) establishes the difference between the intelligence of children whose parents emphasize intellectual achievement and children whose parents place little emphasis on this type of achievement.   Baldwin et al. (1945) also establish that children whose parents were both democratic and warm in their relation with children developed higher levels of intelligence than children whose parents were authoritarian and cold in  their relation with children.

Wolf (1963) has attempted a systematic study of the relation between environmental process variables and general intelligence. Starting from the literature on the effects of environments on intelligence, he hypothesized that 13 process variables could be used to describe the interactions between parents and children insofar as intelligence development is concerned.

A. Press for Achievement Motivation
   1. Nature of intellectual expectations of child.
   2. Nature of intellectual aspirations for child.
   3. Amount of information about child's intellectual development.
   4. Nature of rewards for intellectual development.
B. Press for Language Development
   5. Emphasis on use of language in a variety of situations.
   6. Opportunities provided for enlarging vocabulary.
   7. Emphasis on correctness of usage.
   8. Quality of language models available.
C. Provision for General Learning
   9. Opportunities provided for learning in the home.
   10. Opportunities provided for learning outside the home (excluding school).
   11. Availability of learning supplies.
   12. Availability of books (including reference works), periodicals, and library facilities.
   13. Nature and amount of assistance provided to facilitate learning in a variety of situations.

He devised an interview form with about 60 questions to secure evidence on these variables.   He interviewed the mothers of 60 fifth-grade students in a medium-sized midwestern community. He rated each family or home on each of the 13 process variables, doing all families on one process variable, then on another, etc. He then secured the Henmon-Nelson I.Q.'s on these pupils from the school records. He found a multiple correlation of +.76 between these ratings and the I.Q. This may be contrasted with the correlations of +.40 or less between intelligence and such environmental variables as social status, parent's occupation, or parent's education.

Wolf is still in the process of analyzing the data in his investigation. He believes that some of his environmental process variables (1, 2, 3, 4, 9, 10, 13) are indicative of the parents response to the child and his abilities, whereas other variables (5, 6, 7, 8, 11,12) are likely to be more stable and to represent characteristics of the parents and home which are independent of the child's characteristics and responses. It is significant that both sets of variables are equally correlated with the child's I.Q. (+.70). The lower correlation (as contrasted with the +.76) is attributable to the lower reliability and variability of the shorter scales which include only 6 or 7 of the process variables.

If further research supports Wolf's findings, it will become possible to analyze the ways in which an environment can have a relatively direct influence on general intelligence and various kinds of experimental research on this problem should become possible. Furthermore, it should be possible to develop further Wolf's procedures to determine the maximum effect of the environment on the development of general intelligence. Wolf's study is an illustration of research on a normal range of home environments rather than of extreme conditions in the homes. As such, it represents a considerable advance over previous studies in this area.

Much more research is needed to establish the differences between abundant and deprived environments for the development of general intelligence. The research to date is suggestive of some of the differences, and at least for the extremes in environment there are clear-cut differences in the levels of intelligence reached by children. We do not as yet have a true scale for the measurement of environments in relation to intelligence development, but the available results suggest that when environments are measured with some precision, we should be able to account for some of the variations in individual's intelligence. As intelligence is now measured we believe that the equation

$$MI = f(GP + E)$$

is likely to account for the intelligence test scores at any age, where

MI = measured intelligence, GP = genetic potential, and E = environment.* Furthermore, the relation between parallel measurements of intelligence at any two ages is likely to approach unity if the E (environment) is controlled or accounted for in a statistical relationship. Thus in the Harvard Growth Study the correlation between intelligence at ages 7.4 and 16.4 is found to be +.58, whereas the correlation between intelligence scores at ages 7.4 and 16.4 is found to be +.92 (+1.00 attenuated) when the educational level of extreme groups of parents is included in the multiple correlation

$$R_{\text{I.Q. age 16}} :: \text{I.Q.}_{\text{age 7}} : \text{Environment}_{\text{ages 7—16}}$$

## AFTER AGE EIGHTEEN

Our analyses of intelligence test data so far have been restricted to the first 18 years of life. In most of the analyses of the development of intelligence in terms of an absolute scale it is assumed that intelligence remains constant after about age 20. However, it is likely that when these analyses were made (before 1930) there was very little evidence available on adult intelligence.

The development of the Wechsler Adult Intelligence Test and the adult scales in the 1937 revision of the Stanford-Binet have increased the possibility of securing measurements at later ages. The development of the Army Alpha and the AGCT (Army General Classification Test) have also made possible longitudinal as well as cross-sectional studies of different age groups.

Wechsler (1955) in his standardization of the WAIS (Wechsler Adult Intelligence Scale) tested cross-sectional samples of adults from ages 17 to 75+. Using his verbal score he finds increases until age 30 and then a gradual dropping off of the performance to age 60. Anastasi (1956) points out that the educational levels of Wechsler's samples at various ages are not comparable and that his younger groups have more education than his older groups. Bayley (1955) combined evidence from three longitudinal studies and concluded that there is

---

* The measured intelligence will contain an error term resulting from the unreliability of the measurements. This error can be estimated by the standard error of measurement or it can be reduced by grouping of the data from several measurements. For our purposes, this error term need not concern us at this time. It is also possible that the relation between GP and E is not a simple linear function but it is a more complex interactive function.

continual growth at least until age 50. Jones and Conrad (1933) tested samples at selected ages from 10 to 60 and concluded that general intelligence as measured by the Army Alpha continued to increase from 10 to 22 and then declined slightly until age 60.

If intelligence reached its full development by age 18 and then remained constant until senility, we would expect the correlations after age 18 to be unity or as near unity as the reliability of the measurements would permit. Thus we do know that full stature is attained by age 18 or 20 for most persons. We would be surprised if the correlation between height at age 18 and height at age 40 or 50 was other than +.99. Our knowledge of the developmental curve for stature permits us to predict that height will remain constant after about age 18 and that environmental forces can do almost nothing to change height between ages 20 and 40. We do not subscribe to the thesis that intelligence is a physical or neurological growth function analogous to height growth and that it must have a definite terminal growth point. However, we do know that intelligence as at present measured does reach a virtual plateau in the period ages 10 to 17 and that further development is likely only if powerful forces in the environment encourage further growth and development (Lorge, 1945).

In Table 3.7 we have summarized the longitudinal studies in which the same test was used in repeated measurements. We have compared the observed correlations in each study with the expected correlations from the Overlap Hypothesis based on Bayley's (1955) curve for the development of adult intelligence. Since the Overlap Hypothesis yields estimates for a perfectly reliable test, we have determined the expected correlation reduced by the reliability reported for each test. In almost every instance this reduced theoretical value is higher than the observed correlation. It is also evident that the variability of the college populations for which we do have retest data is considerably smaller than the variability of the general population on which the theoretical estimates are based. We estimate that the standard deviation of intelligence in the general population is about 17 I.Q. points (Terman and Merrill, 1937), whereas the standard deviation of college populations varies from 5.8 to 10.6 I.Q. points (Plant and Richardson, 1958). Rather than try to estimate the standard deviation for each sample in Table 3.7, we have used 11 I.Q. points as the approximate standard deviation for college populations. When Kelley's (1924) formula for the effect of reduced variability on the correlation is used, the expected correlations (corrected for both unreliability and reduced variability) are shown in the last column of Table 3.7. It will be seen that in the 24 compari-

Table 3.7. Observed Correlations Found in Longitudinal Studies of Adult Intelligence
Compared with Theoretically Expected Values

| | | | | | | Correlations | | | |
|---|---|---|---|---|---|---|---|---|---|
| | | | | | | Observed | Theoretically Expected | | |
| | | | | | | A | B | C | D |
| Author and Date | Sample | Test | N | Ages | Gain† | | When Test Reliability Is Perfect | Reduced by Actual Test Reliability | Reduced to Account for Both Reliability and Estimated Variability of College Samples |
| **College Samples** | | | | | | | | | |
| Bailey and Brammer (1951) | Sacramento Junior College | ACE Psy. | 50 | 18–21 | .8* | .78 | .96 | .91 | .82 |
| | College | | 61 | 18–22 | .8* | .78 | .95 | .90 | .79 |
| Barnes (1943) | Univ. of Illinois | ACE Psy. | 105 | 18–20 | | .78 | .96 | .91 | .82 |
| Flory (1940) | College students | ACE Psy. | 74 | 18–22 | | .82 | .95 | .90 | .79 |
| Hunter (1942) | Converse College | ACE Psy. | 43 | 18–22 | 1.1* | .83 | .95 | .90 | .79 |
| | | | 54 | 18–21 | 1.1* | .84 | .96 | .91 | .82 |
| | | | 87 | 18–20 | .9* | .85 | .96 | .91 | .82 |
| Livesay (1939) | College students | ACE Psy. | 50 | 18–22 | .9* | .88 | .95 | .90 | .79 |

| | | | | | | | | | |
|---|---|---|---|---|---|---|---|---|---|
| McConnell (1934) | Cornell College | ACE Psy. | 70 | 18–22 | .9* | .83 | .95 | .90 | .79 |
| Sister Louise (1947) | Marygrove College | ACE Psy. | 86 | 18–21+ | .8* | .82 | .95 | .90 | .79 |
| | | | 107 | 18–21+ | 1.1* | .80 | .95 | .90 | .79 |
| | | | 95 | 18–21+ | .7* | .90** | .95 | .90 | .79 |
| Owens (1953) | Iowa State—adult | Army Alpha | 127 | 19–49 | .6* | .77 | .92 | .89 | .79 |
| Rogers (1930) | Bryn Mawr Study | Thorndike Intell. | 55 | 18–19 | .3 | .63 | .98 | .83 | .70 |
| | | | 35 | 18–20 | .4* | .71 | .96 | .82 | .68 |
| | | | 56 | 18–21 | .5* | .74 | .96 | .82 | .68 |
| Sorenson (1955) | Univ. of Utah | GATB Gen. | 146 | 18–22 | .3* | .77** | .95 | .80 | .66 |
| Tozer (1958) | Liverpool Univ. | NIIP Test | 124 | 18–21+ | 1.0* | .76 | .95 | .90 | .79 |
| **Other Samples** | | | | | | | | | |
| Charles (1953) | Retarded adults | Stanford-Binet | 20 | 27–42 | | .70 | .98 | .93 | .83‡ |
| Terman (1959) | Gifted group | Concept mastery | 768 | 30–42 | 1.3* | .87 | .98 | .93 | .83‡ |
| | Spouses of gifted | | 334 | 30–42 | .9* | .92** | .98 | .93 | .85‡ |
| Bayley (1957) | Berkeley Growth Study | WISC | 33 | 16–18 | | .96 | .96 | .93 | .93‡ |
| | | | 33 | 16–21 | | .95* | .92 | .89 | .89‡ |
| | | | 33 | 18–21 | | .95 | .96 | .93 | .93‡ |

* Significant at .05 level (from value in column D).

** Significant at .01 level (from value in column D).

† Gains are expressed in terms of standard deviation units. Gains significantly different from zero at the .05 level or better are indicated by *.

‡ Adjusted to account for the actual variability of the sample.

sons of observed (column A) and expected correlations (column D), only 4 are significantly different.  We are of the opinion that Bayley's curve for the development of intelligence in adults yields a relatively close approximation to the observed results in the 14 longitudinal studies summarized in Table 3.7.

Undoubtedly, the environments in which people live after age 18 determine the nature of further intellectual development.  Although the extremes of abundance and deprivation are less likely to be found after age 18, at least in samples studied by longitudinal methods, it is likely that some environments are more conducive to further intellectual growth, whereas other environments are likely to maintain but not encourage further growth.  We can even imagine types of environments which would bring about some decline in intelligence.

### STATIC ENVIRONMENTS

The lack of further education after age 18 accompanied by employment in an occupation which makes minimal demands on verbal skills and problem solving would in all likelihood constitute an environment which is not favorable for further development of intelligence.  Such an environment, of course, need not bring about any decline in intelligence.  Thus we might expect the test performance at age 18 and again some years later to be about the same in raw score terms.  However, if age norms are used, it is possible that the derived score at the later age would be somewhat lower than the derived score at the earlier age—not because the individual had lost any of his earlier performance, but because he had not developed further in the meantime.

### ENVIRONMENTS FOR FURTHER GROWTH IN INTELLIGENCE

Higher education with its emphasis on verbal skills and problem solving should constitute a favorable environment for further development of intelligence.  In the eleven studies of general scholastic aptitude of college students presented in Table 3.7 most show a significant amount of growth between ages 18 and 22 (see Gain column).  However, in almost all the studies where the breakdown is possible, the major growth appears to take place in the age period 18 to 20 with only a few studies showing significant growth after age 20.

Although we do not have studies bearing on this matter, we do imagine that certain types of occupations and professions could provide favorable conditions for development of intelligence over long periods of time, whereas other more routine occupations might provide less favorable conditions for further development.

## ENVIRONMENTS FOR DETERIORATION OF ADULT INTELLIGENCE

We find it hard to conceive of environments powerful enough to bring about an actual deterioration of adult intelligence. Such an environment would have to discourage and penalize the individual for utilizing the skills and problem solving abilities he had already developed. Such environments would probably have to affect the individual's personality and morale sufficiently not only to affect his test performance but also his ability to use his verbal skills and problem solving abilities.

Such an extreme environment could conceivably be found in a penal institution where the individual is badly treated, kept for long periods at routine manual work, and not permitted to have books, newspapers, or other forms of communication with the larger world.

Perhaps an approximation to the effects of a deprived environment would be surgery or some form of medical ailment and treatment which brings about a deterioration in personality and destruction of intellectual capabilities (Wittenborn and Mettler, 1951).

We should expect to find only rare instances of such deprived environments for adults since we would expect that an adult who has experienced more stimulating environments and who is free to act will avoid deprived environments or attempt to get out of them as rapidly as possible. It should require considerable coercion to hold an adult in a deprived environment for any length of time.

In general, we would expect an individual to seek to maintain constancy of the environment and to exert considerable effort to avoid or to get out of environments which are significantly different in major respects from environments in which he has managed to maintain himself for a long period of time. Under some circumstances we would expect the individual to seek more complex and challenging environments, especially where the individual is highly motivated and is confident of his ability to handle difficult new problems.

## SPECIFIC TYPES OF MENTAL ABILITIES

Throughout this chapter we have dealt primarily with general intelligence as measured by individual and group intelligence or scholastic aptitude tests. With the development of factor analytic methods

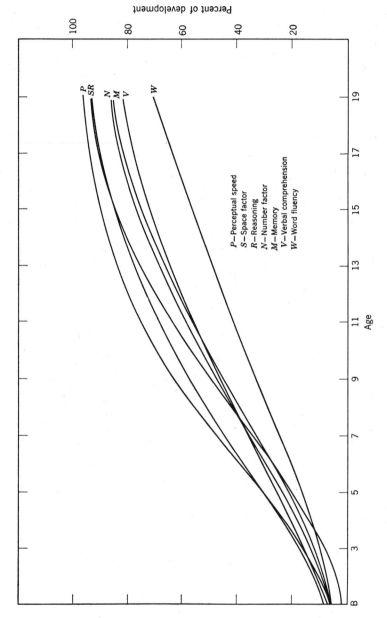

Chart 8. *Estimated Curves for the Development of Specific Mental Abilities* (*Thurstone, 1955*).

P—Perceptual speed
S—Space factor
R—Reasoning
N—Number factor
M—Memory
V—Verbal comprehension
W—Word fluency

Percent of development

Age

increased attention has been given to more specific types of mental abilities.   It is unlikely that each of these more specific mental abilities develops in exactly the same way as the composite or average of these abilities or in exactly the way defined by Thorndike's absolute curve for the development of general intelligence.

There have been a number of longitudinal studies on specific abilities.   These studies have followed groups for one to four years, but none has been concerned with longer periods of time.   That is, we do not have longitudinal studies on specific abilities which have followed the same children from ages 2 or 3 to ages 17 or 18.   However, Thurstone (1955) using cross-sectional data on the Primary Mental Abilities tests has attempted to describe an absolute scale for these abilities.   He made use of data for large groups of children in the age span 5 to 19.   He fitted Gompertz equations to three parameters based on these data.   One is the adult level which he describes as the asymptote toward which the average performance approaches with increasing age.   Another is the zero point which is the point in the scale at which the dispersion of performance vanishes.   The third parameter describes the relative rate at which the adult asymptote is approached with increasing age.   Thurstone's curves for the seven primary abilities corrected to a common scale are shown in Chart 8.   Thurstone uses 80% of the adult performance as one index to compare the different abilities.   He finds that the Perceptual Speed factor reaches 80% of adult performance at age 12, whereas the Space and Reasoning factors attain this level at age 14.   The Number and Memory factors reach this level at about age 16, whereas the Verbal Comprehension factor reaches this level two years later, at age 18.   He reports that Word Fluency reaches the 80% point later than age 20.   We find Thurstone's curves suggestive of differential rates of development for different mental abilities.   It is to be hoped that in the future, there will be more research on methods of developing absolute scales for specific mental abilities and that longitudinal research on these abilities may be related to the absolute scales in the way in which this has already been done for general intelligence.

SUMMARY AND IMPLICATIONS

The stability of intelligence test scores has been a concern of many investigators.   General intelligence is regarded as a very basic and useful measurement of the individual.   However, the educator, clinician, guidance worker, social worker, etc., cannot make long-term decisions about the individual or give help based on observed intelligence test scores unless he can assume some degree of stability of general intelligence.    Many longitudinal studies have been done with measures of general intelligence and scholastic aptitude.    These studies all demonstrate that stability is greater for shorter time periods than for longer time periods.    In addition, the studies reveal increased stability with increased age.    When a number of longitudinal studies are compared with each other and allowances are made for the reliability of the instruments and the variability of the samples, a single pattern clearly emerges.

This general pattern of relationships approximates the absolute scales of intelligence development formulated by Thorndike (1927), Thurstone (1928), and Heinis (1924).    The Thorndike absolute scale fits the longitudinal data most closely during ages 1 to 17.    Both the correlational data and the absolute scale of intelligence development make it clear that intelligence is a developing function and that the stability of measured intelligence increases with age.    Both types of data suggest that in terms of intelligence measured at age 17, about 50% of the development takes place between conception and age 4, about 30% between ages 4 and 8, and about 20% between ages 8 and 17.

These results make it clear that a single early measure of general intelligence cannot be the basis for a long-term decision about an individual.    These results also reveal the changing rate at which intelligence develops, since as much of the development takes place in the first 4 years of life as in the next 13 years.

There is little doubt that intelligence development is in part a function of the environment in which the individual lives.    The evidence from studies of identical twins reared separately and reared together as well as from longitudinal studies in which the characteristics of the environments are studied in relation to changes in intelligence test scores indicate that the level of measured general intelligence is partially determined by the nature of the environment.    The evidence so far available suggests that extreme environments may be described as *abundant* or *deprived* for the development of intelligence in terms of the opportunities for learning verbal and language behavior, opportunities for direct as well as vicarious experience with a complex world, encouragement of problem solving and independent thinking, and the types of expectations and motivations for intellectual growth.

The effects of the environments, especially of the extreme environments, appear to be greatest in the early (and more rapid) periods of intelligence development and least in the later (and less rapid) periods of development. Although there is relatively little evidence of the effects of changing the environment on the changes in intelligence, the evidence so far available suggests that marked changes in the environment in the early years can produce greater changes in intelligence than will equally marked changes in the environment at later periods of development.

Much more research is needed to develop precise descriptions and quantitative measurements of environments as they relate to the development of intelligence. More research is also needed, especially of a longitudinal nature, on the amount of change in intelligence which can be produced by shifting a person from one environment to another. However, a conservative estimate of the effect of extreme environments on intelligence is about 20 I.Q. points. This could mean the difference between a life in an institution for the feeble-minded or a productive life in society. It could mean the difference between a professional career and an occupation which is at the semi-skilled or unskilled level. A society which places great emphasis on verbal learning and rational problem solving and which greatly needs highly skilled and well-trained individuals to carry on political-social-economic functions in an increasingly complex world cannot ignore the enormous consequences of deprivation as it affects the development of general intelligence. Increased research is needed to determine the precise consequences of the environment for general intelligence. However, even with the relatively crude data already available, the implications for public education and social policy are fairly clear. Where significantly lower intelligence can be clearly attributed to the effects of environmental deprivations, steps must be taken to ameliorate these conditions as early in the individual's development as education and other social forces can be utilized.

Although we have made use of measurements at age 17 as the criterion for the analysis of the development of intelligence, there is considerable evidence that intelligence may continue to develop into the adult years. Bayley's curve of the development of adult intelligence does, in general, fit much of the longitudinal evidence on the adult years. This curve shows continued but slow development until age 50. It is quite possible that the shape of the growth curve of intelligence after age 17 is more a function of the environments in which individuals live and work than it is a consequence of biological and maturational processes. This view is supported by longitudinal studies of persons in different occupations and of persons who receive varying amounts of education after age 16. The effect of environment on general intelligence is also demonstrated by the significant increases in measured intelligence during the first year of college in contrast with the smaller increments over the next three years of college, suggesting that new and intensive learning experiences have a more powerful effect than the continuation of these same experiences.

Much of our concern in this chapter is with general intelligence as measured by individual and group tests. However, intelligence may also be broken down into more specific types of abilities. Each of these specific abilities has its own specific developmental curve. Thurstone (1955) has demonstrated that there is a differential growth rate for selected mental abilities with perceptual, space, and reasoning abilities developing somewhat earlier than numerical and memory abilities. Verbal factors, according to Thurstone, develop more slowly. Guidance and curriculum specialists will need more careful and precise descriptions of the development of many specific aptitudes as well as an understanding of the particular environmental conditions which affect the development of these functions.

Our attempts to describe the development of intelligence have been really attempts to describe stability and change in measurements of intelligence. Such measurements are based on particular tests and test problems, and these measurements are undoubtedly affected by the experiences individuals have had both in school and out of school. It seems likely that performance on these tests is responsive to the experiences individuals have had and that change in the general picture of stability and change could be produced by new developments in education and by different child-rearing practices. All this is merely an attempt to alert the reader to the view that our picture of stability and change in measured intelligence is one based on things as they now are, and this includes the particular tests to measure intelligence, the child-rearing practices of families in Western cultures, and educational practices in the schools. It is conceivable that changes in any or all of these could produce a very different picture than the one we have been able to draw. It is to be hoped that we can find ways of prolonging the growth of general intelligence throughout life. It is to be hoped that we can drastically reduce the incidence of low levels of intelligence and increase the proportion of individuals reaching high levels of measured intelligence. There is some evidence that the secular trend in the increase of height over the past 40 years is paralleled by a similar trend in the increase of general intelligence over several decades (Terman and Merrill, 1937). It will be useful for behavioral scientists to understand just why such a trend develops and how it has been influenced by various conditions in the society. Our present picture should serve as a point of departure rather than as a picture of the "natural conditions" or "actual limitations" which determine stability and change in general intelligence.

REFERENCES

Anastasi, A., 1956. Age changes in adult test performance. *Psychol. Reports*, **2**, 509.

Anastasi, A., 1958. Differential Psychology. New York: Macmillan.

Anastasi, A., 1958. Heredity, environment, and the question "How?" *Psychol. Review*, **65**, 197–208.

Anderson, J. E., 1939. The limitations of infant and pre-school tests in the measurement of intelligence. *J. Psychol.*, **8**, 351–379.

Anderson, L. D., 1939. The predictive value of infant tests in relation to intelligence at five years. *Child Develpm.*, **10**, 203–212.

Bailey, D. W., and Brammer, L. M., 1951. Variability in A. C. P. E. scores from freshman to junior and senior years. *Calif. J. Ed. Res.*, **2**, 159–164.

Baldwin, A. L., Kalhorn, J., and Breese, F. H., 1945. Patterns of parent behavior. *Psychol. Monogr.* 58, 3, (Whole No. 268).

Barnes, M. W., 1943. Gains on the A. C. E. Psychological Examination during the freshman-sophomore years. *School and Society*, **57**, 250–252.

Bayley, N., 1949. Consistency and variability in the growth of intelligence from birth to eighteen years. *J. Gen. Psychol.*, **75**, 165–196.

Bayley, N., 1954. Some increasing parent-child similarities during the growth of children. *J. Educ. Psychol.*, **45**, 1–21.

Bayley, N., 1955. On the growth of intelligence. *Amer. Psychol.*, **10**, 805–818.

Bayley, N., 1957. Data on the growth of intelligence between sixteen and twenty-one years as measured by the Wechsler-Bellevue scale. *J. Gen. Psychol.*, **90**, 3–15.

Bernstein, B., 1960. Aspects of language and learning in the genesis of the social process. *J. Child Psychol. and Psychiatry (Great Britain)*, **1**, 313–324.

Bradway, K. P., 1944. I.Q. constancy in the revised Stanford-Binet from the preschool to the junior high school level. *J. Gen. Psychol.*, **65**, 197–217.

Bradway, K. P., 1945. Predictive value of Stanford-Binet pre-school items. *J. Ed. Psychol.*, **36**, 1–16.

Burks, B. S., 1928. The relative influence of nature and nurture upon mental development. *Nat. Soc. for the Study of Ed. Yearbook*, **27**, I, 219–316.

Burks, B. S., Jensen, D. W., Terman, L., et al., 1930. Genetic studies of genius. Vol. III, The promise of youth. Stanford, California: Stanford Univ. Press.

Burt, Sir, C. L., 1958. The inheritance of mental ability, *American Psychol.*, **13**, 1–15.

Cattel, P., 1931. Constant changes in the Stanford-Binet I.Q., *J. Ed. Psychol.*, **22**, 544–550.

Cavanaugh, M. C., et al., 1957. Prediction from the Cattel Infant Intelligence Scale. *J. Consult. Psychol.*, **21**, 33–37.

Charles, D. C., 1953. Ability and accomplishment of persons earlier judged mentally deficient. *Gen. Psychol. Monogr.*, **47**, 3–71.

Cronbach, L. J., 1960. Essentials of psychological testing. New York: Harper.

Dennis, W., and Najarian, P., 1957. Infant development under environmental handicap. *Psychol. Monogr.*, **71**, No. 7, Whole No. 463.

Ebert, E., and Simmons, K., 1943. The Brush Foundation study of child growth and development. I. Psychometric tests. *Monogr. Soc. Res. Child Develpm.* **8**, No. 2, 1–113.

Eels, K., et al. 1951.   Intelligence and cultural differences.   Chicago: Univ. of Chicago Press.

Flory, C. D., 1940.   The intellectual growth of college students.   *J. Ed. Res.*, **33**, 443–451.

Freeman, F. N., and Flory, C. D., 1937.   Growth in intellectual ability as measured by repeated tests.   *Monogr. Soc. Res. Child Develpm.*, **2**, No. 2.

Freeman, F. N., Holzinger, K. J., and Mitchell, B. C., 1928.   The influence of environment on the intelligence, school achievement, and conduct of foster children.   *Nat. Soc. for the Study of Ed. Yearbook*, Vol. 27, I, 103–218.

Goldin M. R., and Rothschild, S., 1942.   Stability of intelligence quotients of metropolitan children of foreign-born parentage.   *Elementary School J.*, **42**, 673–676.

Guilford, J. P., and Michael, W. B., 1948.   Approaches to univocal factors.   *Psychometrika*, **13**, 1–22.

Havighurst, R. J., and Janke, L. L., 1944.   Relation between ability and social status in a midwestern community.   I. Ten-year old children.   *J. Ed. Psychol.*, **35**, 357–368.

Heinis, H., 1924.   La loi du developpement mental.   *Archives de Psychologie*, **74**, 97–128.

Hilden, A. H., 1949.   A longitudinal study of intellectual development.   *J. Psychol.*, **28**, 187–214.

Hirsch, N. D. M., 1930.   An experimental study upon 300 school children over a six year period.   *Gen. Psychol. Monogr.*, **7**, 487–549.

Hofstaetter, P. R., 1954.   The changing composition of "intelligence:" a study in T-technique.   *J. Genet. Psychol.*, **85**, 159–164.

Honzik, M. P., 1957.   Developmental studies of parent-child resemblance in intelligence.   *Child Develpm.*, **28**, 215–228.

Honzik, M. P., Macfarlane, J. W., and Allen, L., 1948.   The stability of mental test performance between two and eighteen years.   *J. Experimental Ed.*, **17**, 309–324.

Hunt, J. McV., 1961.   Intelligence and experience.   New York: Ronald Press.

Hunter, E. C., 1942.   Changes in scores of college students on the American Council Psychological Examination at yearly intervals during the college course.   *J. Ed. Res.*, **36**, 284–291.

Husén, T., 1959.   Psychological twin research.   Stockholm: Almquist and Wicksell.

Jones, H. E. and Conrad, H. S., 1933.   The growth and decline of intelligence: A study of a homogeneous group between the ages of ten and sixty.   *Gen. Psychol. Monogr.*, **13**, 223–294.

Katz, E., 1941.   The constancy of the Stanford-Binet I.Q. from three to five years.   *J. Psychol.*, **12**, 159–181.

Kelley, T. L., 1924.   Statistical method.   New York: Macmillan.

Kirk, S. A., 1958.   Early education of the mentally retarded.   Urbana: Univ. of Illinois Press.

Klineberg, O., 1931.   A study of psychological differences between "racial" and national groups in Europe.   *Archives Psychol.*, **132**.

Knezevich, S., 1946.   The constancy of the I.Q. of the secondary school pupil.   *J. Ed. Res.*, **39**, 506–516.

Layton, W. L., 1954.   The relation of ninth grade test scores to twelfth grade test scores and high school rank.   *J. Appl. Psychol.*, **38**, 10–11.

Leahy, A. M., 1935.   Nature-nurture and intelligence.   *Gen. Psychol. Monogr.*, **17**, 235–308.

Lee, E. S., 1951.   Negro intelligence and selective migration: A Philadelphia test of the Klineberg hypothesis.   *Am. Sociol. Rev.*, **16**, 227–233.

Levy, R. J., 1947. Effects of institutional versus boarding home care on a group of infants. *J. Personality*, **15**, 233–241.

Livesay, T. M., 1939. Does test intelligence increase at the college level? *J. Ed. Psychol.*, **30**, 63–68.

Loevinger, J., 1943. On the proportional contributions of differences in nature and in nurture to differences in intelligence. *Psychol. Bulletin*, **40**, 725–756.

Lorge, I., 1945. Schooling makes a difference. *Tchrs. Coll. Record*, **46**, 483–92.

Louise, Sister M. F., 1947. Mental growth and development at the college level. *J. Ed. Psychol.*, **38**, 65–82.

Maurer, K. M., 1946. Intellectual status at maturity as a criterion for selecting items in pre-school tests. Minneapolis: Univ. of Minnesota Press.

McConnell, T. R., 1934. Changes in scores in the psychological examination of the American Council on Education from freshman to senior year. *J. Ed. Psychol.*, **25**, 66–69.

McNemar, Q., 1942. The revision of the Stanford-Binet scale. Boston: Houghton Mifflin.

McNemar, Q., 1949. Psychological statistics. New York: Wiley.

Milner, E. A., 1951. A study of the relationship between reading readiness in grade one school children and patterns of parent-child interaction. *Child Develpm.*, **22**, 95–112.

Newman, H. H., Freeman, F. N., and Holzinger, K. J., 1937. Twins: a study of heredity and environment. Chicago: Univ. of Chicago Press.

Owens, W. A., 1953. Age and mental abilities: a longitudinal study. *Gen. Psychol. Monogr.*, **48**, 3–54.

Pintner, R., and Stanton, M., 1937. Repeated tests with the C. A. V. D. scale. *J. Ed. Psychol.*, **28**, 494–500.

Plant, W. T., and Richardson, H., 1958. The I.Q. of the average college student. *J. Counsel. Psychol.*, **5**, 229–231.

Roff, M. E., 1941. A statistical study of the development of intelligence test performance. *J. Psychol.*, **11**, 371–386.

Rogers, A. I., 1930. The growth of intelligence at the college level. *School and Society*, **31**, 693–699.

Rosca, A., 1939. Inteligenta in mediul Rura-Urban. *Rev. Psihol.*, **2**, 131–141.

Sherman, M., and Key, C. B., 1932. The intelligence of isolated mountain children. *Child Develpm.*, **3**, 279–290.

Skodak, M., and Skeels, H. M., 1945. A follow up study of children in adoptive homes. *J. Gen. Psychol.*, **66**, 21–58.

Skodak, M., and Skeels, H. M., 1949. A final follow up study of 100 adoptive children. *J. Gen. Psychol.*, **75**, 85–125.

Sontag, L., Baker, C., and Nelson, V., 1958. Mental growth and personality: a longitudinal study. *Monogr. Soc. Res. Child Develpm.*, **23**, No. 2, 1–143.

Sorenson, G., and Senior, N., 1955. Changes in G. A. T. B. scores with college training. *Calif. J. Ed. Res.*, **6**, 170–173.

Spitz, R. A., 1945. Hospitalism. An inquiry into the genesis of psychiatric conditions in early childhood. *Psychoanal. Stud. Child.*, **1**, 53–74.

Stalnaker, R. C., and Stalnaker, J. M., 1946. The effect on a candidate's score of repeating the scholastic aptitude test of the College Entrance Examination Board. *Ed. and Psychol. Measurement.*, **6**, 495–503.

Terman, L. M., and Merrill, M. A., 1937. Measuring intelligence. New York: Houghton Mifflin.

Terman, L. M., and Oden, M. H., 1959. The gifted group at mid-life. Stanford, California: Stanford Univ. Press.

Thorndike, E. L., 1927. The measurement of intelligence. New York: Teachers College, Columbia University.

Thorndike, R. L., 1940. Constancy of the I.Q. *Psychol. Bulletin*, **37**, 167–186.

Thurstone, L. L., 1928. The absolute zero in intelligence measurement. *Psychol. Rev.*, **35**, 175–197.

Thurstone, L. L., 1955. The differential growth of mental abilities. Chapel Hill, North Carolina: Univ. of North Carolina Psychometric Laboratory, No. 14.

Townsend, A., 1944. Some aspects of testing in the primary grades. *Ed. Records Bulletin*, No. 40, 51–54.

Tozer, A. H. D. and Larwood, H. J. C., 1958. The changes in intelligence test scores of students between the beginning and end of their university courses. *Brit. J. Ed. Psychol.*, **28**, 120–128.

Traxler, A., 1934. Reliability, constancy and validity of the Otis I.Q. *J. Appl. Psychol.*, **18**, 241–251.

Wechsler, D., 1939. Measurement of adult intelligence. Baltimore: Williams and Wilkins.

Wechsler, D., 1955. Wechsler adult intelligence scale, manual. New York: Psychological Corporation.

Wellmen, B. L., 1937–1938. Mental growth from pre-school to college. *J. Exp. Ed.*, **6**, 121–138.

Wentworth, M. M., 1926. Individual differences in the intelligence of school children. Harvard Studies in Education No. 7. Cambridge: Harvard Univ. Press.

Wheeler, L. R., 1932. The intelligence of East Tennessee mountain children. *J. Ed. Psychol.*, **23**, 351–370.

Wheeler, L. R., 1942. A comparative study of the intelligence of East Tennessee mountain children. *J. Ed. Psychol.*, **33**, 321–334.

Winons, J. M., 1949. Changes in I.Q. and M. A. between the 8th and 12th grade levels. Unpublished Ph.D. Dissertation, Univ. of California.

Wittenborn, J. R., and Mettler, F. A., 1951. Some psychological changes following psychosurgery. *J. Abn. and Soc. Psych.*, **46**, 548–556.

Wolf, R. M., 1963. The identification and measurement of environmental process variables related to intelligence. Ph.D. Dissertation in progress, Univ. of Chicago.

Woodworth, R. S., 1941. Heredity and environment: a critical survey of recently published materials on twins and foster children. New York: *Soc. Sci. Res. Council Bulletin*, No. 47.

*Chapter Four*

# STABILITY OF ACHIEVEMENT DATA

## INTRODUCTION

Our interest in the stability of school achievement was greatly stimulated by a study reported by Geraldine Spaulding (1960) of the Educational Records Bureau. In this study, Spaulding found correlations of approximately +.80 between high school grades and freshman college grades, corrected for the differences in standards between colleges.

Spaulding went one step further and asked whether 12th year grades were better for predicting college grades than 11th year grades, 10th year grades, or 9th year grades. She found that an average based on four years of secondary grades was superior to one year of grades primarily because of the increased reliability of the four year average. She found that there was little superiority for prediction purposes of 12th year grades over the 11th or 10th year grades. Although the 9th year grades were not quite as good as the 12th year grades for prediction of college grades, the difference was relatively small. In other words, Spaulding had found considerable evidence for the stability of grades assigned students. When the correlation of +.68 between 9th year grades and college freshman grades is corrected for attenuation,* the theoretical correlation (assuming perfect reliability of the measures) is approximately +.85.

In our own study of high school and college grades (Bloom and Peters, 1961), we had found correlations as high as +.83 between high school grades and college grades (scaled), with the average correlation being approximately +.78. When these correlations are corrected for

* Assuming a reliability of .80 for one year of grades at the secondary or college level.

the unreliability of grades, the theoretical correlation is +.92. Correlations of this magnitude are extremely rare in educational research and one is inclined to look for some flaw in the research or in the statistical procedures. The most likely source of error is in the correction for attenuation. Perhaps the estimation of the reliability of grades (+.90 for four years of high school grades and +.80 for one year of college grades) is too low.

With this in mind, the writer turned to the research of Learned and Wood (1938) in the Pennsylvania Study. This study was one in which a battery of achievement tests was given to students at the end of high school and then repeated at the end of two and four years of college. Although comparisons between high school and college scores could not be made because some of the tests were altered between the high school and college testing, the results of the tests at the end of the 14th and 16th years of school yielded correlations of +.90. Since the total battery of achievement tests had a reliability of +.98, it was unnecessary to use the correction for attenuation. These results were based on 2830 cases in 45 Pennsylvania colleges. The data in this study were gathered in the period 1928 to 1932. Lannholm (1946) gave the General Education portion of the Graduate Record Examination to over 1000 college sophomore students. He repeated the tests when these students were seniors. In this study, which included students in 16 different colleges, the correlation between sophomore and senior scores was +.91.* Heston (1950) gave the same portion of the Graduate Records Examination to students in a single college in their sophomore year and then again when they were seniors. His results, a correlation of +.90, demonstrates the same stability of achievement in a single college. Thus studies carried on over almost a third of a century (1928 to 1961) yield much the same evidence of stability whether the results are based on grades or achievement tests.

STABILITY OF TEACHER MARKS AND
ACHIEVEMENT TEST SCORES

We have tried to assemble the studies on the stability of achievement indices but found, to our surprise, that no study was available which followed the same children over the entire period of public schooling from grades 1 to 12. We found a large number of studies which

---

* We estimate the reliability of the General Education Index of the Graduate Record Examination to be *over* .95, so there is little need to correct these correlations for attenuation.

followed the same students for periods of one or two years, but only a few studies which followed students for 3 to 8 years. The most prominent of these longer term studies are the five studies briefly described below.

Scannell (1958) gave the Iowa Tests of Basic Skills to students at the 4th, 6th, and 8th grade and the Iowa Tests of Educational Development to these same students when they were in the 9th and 12th grades. The correlation for 581 students between their scores at the 4th and 12th grades was $+.72$ (corrected for attenuation $= +.74$).

Haggerty (1941) followed a group of children from the 6th grade through the 9th grade. He administered the Stanford Achievement Test to children in a single school and found a correlation of $+.89$ between the 6th and 9th grade scores. When he corrected this correlation for attenuation, he found it to be $+.95$. Townsend (1951) gave the Stanford Achievement Test to students in the 5th grade and then again in the 7th grade and found a correlation of $+.84$ between these two sets of scores (corrected for attenuation $= +.91$). These included students in a number of different schools.

Traxler (1950) gave a reading comprehension test to students in four schools at the 7th grade and repeated the testing again at the 12th grade. He found a correlation of $+.77$ (corrected for attenuation $= +.85$).

One other study is that of Kelley (1914) who found a correlation of $+.62$ (corrected for attenuation $= +.81$) between teachers' marks at the 4th grade and at the 9th grade. The correlation between teachers' marks at the 6th and 9th grade (corresponding to Haggerty's interval) was $+.73$ (corrected for attenuation $= +.91$).

These studies all suggest that for periods of three to eight years there is considerable evidence of stability, with the corrected correlations being somewhere between $+.75$ and $+.95$.

We were somewhat dismayed by the lack of longitudinal evidence over longer periods of time, especially so since a great deal of achievement testing has been done in the schools over the past forty years and in view of the fact that teachers' marks have been used in the schools since time immemorial. All the results of such testing or grading are matters of record in school files.

In order to secure longitudinal data comparable to that obtained for physical characteristics and intelligence, Alexander (1961) and Hiremath (1962) obtained data on teachers' marks and test results in reading from grades 1 to 12. These achievement indices for a sample of over 200 students obtained from 10 elementary schools and 3 high schools have been analyzed by these two workers.

In Chart 1, we have presented Alexander's and Hiermath's results with grade 8 (age 14) as the criterion. We have also included in this chart the comparable results from several other studies. It will be noted that the results for an achievement battery, teachers' grades, and reading comprehension vary considerably although all show similar patterns. When the correlations are corrected for the unreliability of the measures, the curves (Chart 1a) become more similar, with only minor differences that are probably attributable to the variations in the homogeneity of the samples.

In Chart 2 we have presented the results of several longitudinal studies in which an achievement battery, reading comprehension, or teachers' grades at various levels are related to similar measures at grade 12. Here again, we find considerable differences in the curves although they all show a characteristic pattern. When we correct the correlations for the unreliability of the measures, the curves (see Chart 2a) are now more similar. The Hicklin (1962) curve is somewhat different from the Traxler (1950) curve, but this may be attributed to the fact that Hicklin used two different reading tests (Chicago Reading Test at grades 4 and 5, and the Cooperative English Test

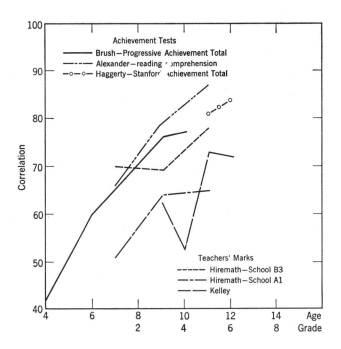

*Chart 1. Correlations between Achievement at Each Grade and Achievement at Grade 8.*

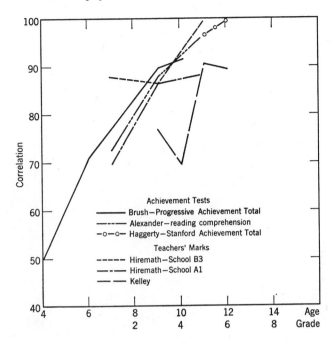

*Chart* 1a. *Correlations between Achievement at Each Grade and Achievement at Grade* 8 *(Corrected for Attenuation).*

at grades 6 to 10) and related these measurements to the CEEB Scholastic Aptitude Test verbal score at grade 11. It is likely that these shifts in the measures would yield lower correlations than those reported by Traxler (1950) who used the Cooperative English Test from grades 7 through 12.

Although we are able to demonstrate considerable stability on both achievement test results and teachers' marks, only in case of teachers' marks do we have longitudinal data for the 12 years from grade 1 to 12. In order to determine whether there is any way of combining several longitudinal studies in such a way that we can secure estimates of the relationship between achievement test results over the 12 years of public school, we have turned to a brief study of the relationship between test scores at one grade and gains in succeeding years.

In Table 4.2 we have summarized the results of a number of studies on the correlations between measures at one age and gains in subsequent intervals. It will be noted that in general the correlations between initial measures and gains in subsequent years are very low. The median of the correlations is −.07 and it is likely that, in general,

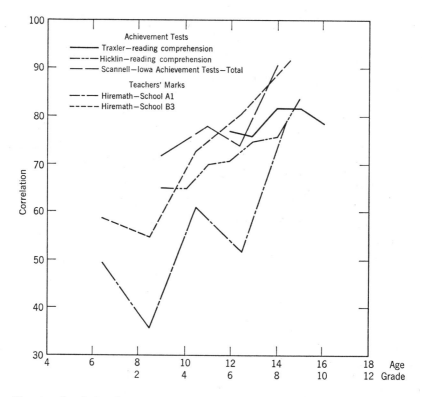

*Chart 2.  Correlations between Achievement at Each Grade and Achievement at Grade 12.*

position at one measurement and gains in the two to six years which follow will show a low relationship. We shall discuss this point in greater detail at a later point in this chapter. Here we wish to make use of this finding as a basis for combining two or more longitudinal studies.

If we assume that the correlation between tests at one grade and the gain from that grade to another grade is zero (see Table 4.2), then we may make use of the following relationships.

1.
$$r_{13.2} = \frac{r_{13} - r_{12} \times r_{23}}{\sqrt{1 - r^2_{12}} \sqrt{1 - r^2_{23}}}$$

where 1 is the initial test, 2 is the second test, and 3 is the third test. If then

2.
$$r_{13.2} = 0, \quad \text{or} \quad r_{13} - r_{12} \times r_{23} = 0$$

then

3.                                      $r_{13} = r_{12} \times r_{23}$

Using this set of relationships, we may then attempt to combine several longitudinal studies.

Alexander (1961) reported the correlations for the Chicago Reading Tests between grades 2 and 8, and Traxler (1950) reported the correlations for the Cooperative English Tests between grades 8 and 12. Although we do recognize that variations in the samples and variations in the tests may affect the results, we have applied formula 3 to estimate the correlations from grades 2 to 12 by combining these two studies. We have plotted the results in Chart 3.

It is unfortunate that we are not able to secure adequate evidence of educational achievement before grade 2 or age 8. The very nature of academic achievement restricts our longitudinal study of school achievement to the period of time the students are in school. How-

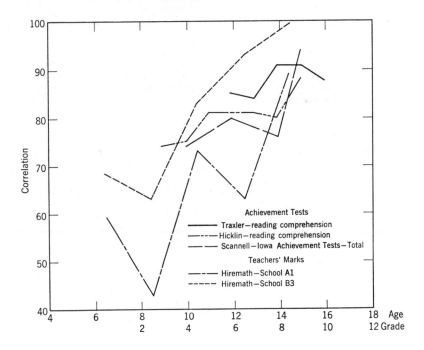

Chart 2a.   *Correlations between Achievement at Each Grade and Achievement at Grade 12 (Corrected for Attenuation).*

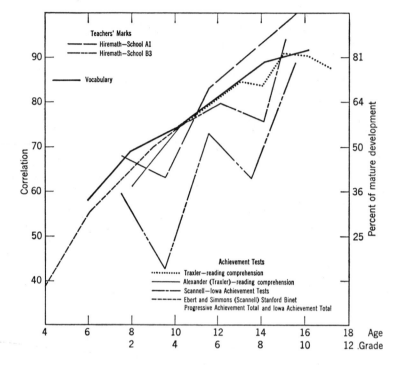

*Chart 3.    Correlations between Achievement at Each Grade and Achievement at Grade 12 (Corrected for Attenuation) in Contrast with the Overlap Hypothesis Based on Vocabulary Development.*

ever, the nature of the curves presented in Chart 3 suggests that learning patterns are partially established before children enter school.

One study that bears on this point was reported by Ebert and Simmons (1943). They gave the Progressive Achievement Tests to students at age 13. These same children had been given intelligence tests at age 4 and at several additional points thereafter. They found the correlation between I.Q. at age 6 and school achievement at age 13 to be +.60. This is a minimal estimate since the intelligence test probably samples a type of learning which is not strictly comparable to school achievement. If we look for an estimate of how high a general intelligence test correlates with an achievement battery when both are taken at the same time (at about age 13), we begin to get an estimate of the comparability of the two kinds of learning. Several studies report such correlations. (Kelley, 1927; Coleman and Cureton, 1954.) These average about +.85. If we use this as a basis for determining the theoretical relationship and divide +.85 into the obtained figures in the Ebert-Simmons study, it is evident that by age 6 the theoretical correlation with school achievement 7 years later is as high as +.71.

Thus, we find that $r_{6-13}$ derived from Eberts and Simmons is +.71, and $r_{13-18}$, using Scannell (1958) is +.77 when both are corrected for attenuation. Substituting these values in formula 3, we estimate the correlation between ages 6 and 18 to be +.55.

Similarly, Ebert and Simmons (1943) report a correlation of +.42 between the Stanford Binet at age 4 and the Progressive Achievement battery at age 13. This would be +.50 (corrected for the correlation between the two tests given at the same time). Substituting this value and the Scannell (1958) value of +.77 for ages 13 to 18 in formula 3, we estimate the correlation between ages 4 and 18 to be +.39. These estimates have been plotted in Chart 3.

The curves in Chart 3 show the general patterns of stability found in the studies we have been able to locate as well as the estimates we have made for ages 4 and 6. It seems reasonable to assume that Anderson's Overlap Hypothesis, which has been used in the chapters on physical characteristics and intelligence, also applies to general scholastic achievement. However, we have been somewhat at a loss as to how to develop or select an absolute scale of general scholastic development for purposes of comparison with our correlational data. General scholastic achievement is an even more general quality than the I.Q. and we could see no satisfactory way of analyzing our general achievement tests to produce an absolute scale.

We were struck by the largely verbal character of the measures of general achievement, and in our search for a general developmental scale that would have relevance for school achievement we reasoned that the development of language and the development of school achievement might be parallel. That is, a developmental scale of language would be likely to exhibit developmental characteristics very similar to those found in our longitudinal studies of general achievement.

Moving from language development to the vocabulary tests seemed an easy transition. What we needed then was a vocabulary test which represented a sampling of the words in a dictionary (rather than of particular word frequencies). If, then, the number of words on the test sample known by the examinee could be related to estimates of the number of words in the dictionary known by the examinee, we would have estimates of the development of total vocabulary which could be used as an approximation to an absolute scale. Although this procedure does make the assumption that each word is the unit and that the difference between knowing 7000 versus 10,000 is the same as the difference between knowing 17,000 versus 20,000 words, or that in terms of words known 5000 is to 10,000 as 10,000 is to 20,000, we are of the opinion that for practical purposes this assumption is not entirely unreasonable.

Seashore and Eckerson (1940), Smith (1941), and Bryan (1953) have attempted to estimate the size of vocabulary known by individuals at different ages and grade levels.   Although these different workers have used different sampling methods, estimates of vocabulary size at different ages in terms of percent of the number of words known at age 18 are quite similar.   The Smith (1941) and the Seashore and Eckerson (1940) estimates seem to us to be most suited for our purposes since they include estimates of the size of the vocabulary of college students as well as those for the earlier years.   We have drawn the curve for vocabulary growth on Chart 3 for purposes of contrast with the correlational data on general scholastic achievement.   With only slight discrepancies, the developmental characteristics of vocabulary fit the correlational curves for reading comprehension, general achievement test scores, as well as teachers' marks.   Although we do not wish to argue that there is a causal relationship between language development and general school learning, we do believe that one is a good reflection of the other and that language development and school learning probably develop in much the same way.

Returning then to Chart 3, we would conclude that both on the basis of longitudinal studies of general achievement, reading comprehension, and teachers' marks and on the basis of general vocabulary development (from cross-sectional studies), the curve for the development of general scholastic achievement exhibits a relatively rapid early development followed by a rather steady rate of growth from ages 9 to 18.

In terms of the Overlap Hypothesis, the half development using age 17 or 18 as the criterion is at about age 8, whereas the two-thirds development is at about age 12.   When the student starts school at age 6, he has attained about one-third of his general learning pattern, whereas by age 4 he is likely to have attained about one-sixth of this pattern.

How well do these estimates account for the findings in other longitudinal studies?   In order to test the accuracy of our general curve we have collected a large number of studies in which the same or parallel tests have been used at intervals of one year or more.   Insofar as possible we have included only those studies in which students in a single school with a "common set of learning experiences" have been studied. Perhaps the point of this is most clearly illustrated by a specific subject like mathematics or physics. If 200 students take an initial test in mathematics, and then a random sample of 100 of these subsequently take a course in mathematics, whereas the other 100 do not, the retest in mathematics for all 200 students is likely to have a low correlation with the initial test, although the correlation between initial and retest for each sample of 100 is likely to be very high. Thus the stability of mathematical achievement may be confounded by the differences in the experiences of the two groups.

We have also restricted our selection to studies in which tests with a reliability of +.85 or higher have been used. Tests with lower reliabilities would include so much error variance that our correlations would not be very meaningful.

It will be noticed in Table 4.1 that our theoretical values represent a very close approximation to almost two-thirds of the correlations reported in the different studies in which longitudinal data on achievement are included. That is, only one-third of the correlations are significantly different from the theoretical values. This appears to be equally true at the elementary and secondary education levels.* It is generally true of the different subjects, and does not differ for tests of knowledge in contrast to tests of the higher mental processes of problem solving.

We may conclude from this that our theoretical values are relatively good approximations of the observed values in a great variety of studies conducted during the past thirty years with achievement tests. Furthermore, data based on teachers' marks reflect much the same general trend.

SUMMARY OF GROWTH

We may conclude from our results on general achievement, reading comprehension, and vocabulary development that by age 9 (grade 3) at least 50% of the general achievement pattern at age 18 (grade 12) has been developed, whereas at least 75% of the pattern has been developed by about age 13 (grade 7). The evidence from the Ebert-Simmons (1943) study as well as studies of vocabulary development

---

* The correlations at the higher education level are very close to the theoretical values, but because they are at the extreme end of the correlations small differences are significant.

Table 4.1. Test-Retest Studies of School Achievement

| | | | | | | Correlations | |
| | | | | | Observed | Theoretically Expected | |
| | | | | | A | B When Test Reliability Is Perfect | C Reduced by Actual Test Reliability |
| Author and Date | Sample | Test | N | Grades | | | |
|---|---|---|---|---|---|---|---|
| Elementary | | | | | | | |
| Hildredth (1936) | School children | Stanford Ach. Tot. | 47 | 2.9–6.9 | .54** | .84 | .79 |
| Townsend (1944) | School children | Reading | 47 | 2.9–6.9 | .67 | .84 | .60 |
| | School children | Met. Ach. Tot. | 105 | 2–5 | .61** | .89 | .77 |
| | School children | Av. Rdg. | 105 | 2–5 | .76 | .89 | .79 |
| | School children | Eng. + Language | 105 | 2–5 | .59** | .89 | .76 |
| Traxler (1941) | School children | New Stanford Ach. Tot. | 81 | 5–6 | .90 | .95 | .89 |
| | School children | Paragraph Meaning | 81 | 5–6 | .77 | .95 | .81 |
| Elementary-secondary | | | | | | | |
| Haggerty (1941) | School children | New Stanford Ach. Tot. | 171 | 6–9 | .89 | .91 | .86 |
| | School children | New Stanford Ach. Tot. | 171 | 7–9 | .93* | .95 | .89 |

| Study | Population | Test | N | Grade | | | |
|---|---|---|---|---|---|---|---|
| Haggerty (1941) | School children | New Stanford Ach. Tot. | 171 | 8–9 | .93 | .98 | .92 |
| Kvaraceus and Lanigan (1948) | School children | ITBS Rdg. | 27 | 6.9–8.9 | .75 | .95 | .78 |
| | School children | Vocab. | 27 | 6.9–8.9 | .74 | .95 | .78 |
| | School children | Language | 27 | 6.9–8.9 | .61 | .95 | .78 |
| Kelley (1914) | School children | Teachers' marks | 174 | 4–9 | .62 | .82 | .66 |
| | School children | Teachers' marks | 174 | 5–9 | .53* | .85 | .66 |
| | School children | Teachers' marks | 174 | 6–9 | .73 | .91 | .73 |
| | School children | Teachers' marks | 174 | 7–9 | .72 | .95 | .74 |
| Townsend (1951) | School children | Stanford Ach. Tot. | 56 | 5–7 | .84 | .91 | .86 |
| Secondary | | | | | | | |
| Aaron (1946) | School children | Coop. Eng. Tot. | 885 | 8–10 | .82** | .97 | .92 |
| Adam (1940) | School children | Coop. Eng. Tot. | 241 | 9–12 | .77** | .91 | .86 |
| Traxler (1950) | School children | Coop. Read. Comp. Tests | 36 | 7–12 | .77 | .86 | .80 |
| | | | 36 | 8–12 | .76 | .90 | .85 |
| | | | 36 | 9–12 | .82 | .91 | .86 |
| | | | 36 | 10–12 | .82 | .92 | .86 |
| Hicklin (1962) | School children | Teachers' marks | 92 | 9–12 | .73 | .91 | .76 |
| | School children | Teachers' marks | 92 | 10–12 | .79 | .92 | .77 |
| | School children | Teachers' marks | 92 | 11–12 | .80 | .96 | .81 |

Table 4.1. Test-Retest Studies of School Achievement (Continued)

Correlations

| | | | | | Observed | Theoretically Expected | |
| | | | | | A | B<br>When Test<br>Reliability Is<br>Perfect | C<br>Reduced by<br>Actual Test<br>Reliability |
|---|---|---|---|---|---|---|---|
| Author and Date | Sample | Test | N | Grades | | | |
| Krantz (1957) | School children | Iowa Basic Skills and Iowa Ed. Develop. | 256 | 7–9 | .83** | .95 | .90 |
| | School children | Iowa Basic Skills and Iowa Ed. Develop. | 251 | 7–11 | .79* | .89 | .85 |
| Secondary and Higher Education or Adult Life | | | | | | | |
| Byrnes and Henmon (1935) | H.S. and Univ. of Wis. students | Grades | 250 | Four yr. H.S. –Four yr. Univ. of Wis. | .74 | .88 | .77 |
| | H.S. and Univ. of Wis. students | Grades | 250 | Four yr. H.S. –1st sem. college | .72 | .91 | .74 |
| | H.S. and Univ. of Wis. students | Grades | 250 | Tenth grade –1st sem. college | .64 | .89 | .67 |

| Study | Group | Test | N | Grade/Age | | | |
|---|---|---|---|---|---|---|---|
| Finch and Nemzeh (1934) | | | 90 | H.S. and college | .79 | .88 | .70 |
| Lorge (1934) | H.S. and Univ. of Minn. students School children followed up 10 years later | Honor Point Average Thorndike McCall Rdg., Scale | 860 | 8–age 24 | .57* | .81 | .62 |
| **Higher Education** | | | | | | | |
| Chausow (1955) | College Students | ACE Critical Thinking Test | 136 | 13–14 | .68** | .95 | .80 |
| Hartmann and Barrick (1934) | College Students | Carnegie Gen. Culture Test | 77 | 14–16 | .75** | .96 | .93 |
| Heston (1950) | College Students | Grad. Record Gen. Ed. Index | 157 | 14–16 | .90* | .96 | .93 |
| Lannholm (1952) | College Students | Grad. Record Gen. Ed. Index | 1012 | 14–16 | .91** | .96 | .93 |
| Learned and Wood (1938) | College students | Carnegie Gen. Culture Test | 2830 | 14–16 | .89** | .96 | .93 |
| Morris (1953) | College students | ACE Critical Thinking Test | 628 | 13–14 | .71** | .95 | .80 |
| Silvey (1951) | College students | Nelson Denny Reading | 517 | 13–14 | .83 | .95 | .86 |

* Significant at .05 level.
** Significant at .01 level.

suggest that about one-third has been developed by the time the individual has entered school.

Putting these data in another form, we would estimate the following values for the development of general learning as based on overall achievement indices (achievement test batteries, reading comprehension, and teachers' marks) or based on general vocabulary development.

|  |  | *Percent of Age* 18 *Development* | |
| --- | --- | --- | --- |
|  |  | *Each Period* | *Cumulative* |
| Ages | Birth–6 | 33 | 33 |
| Ages | 6–13 | 42 | 75 |
| Ages | 13–18 | 25 | 100 |

Although these figures may be debatable and further studies may modify them, we are of the opinion that the relative magnitude of these values will not change appreciably. We believe it is likely that more careful investigations will reveal even larger values for the preschool period and the first three years of elementary school than is suggested by the studies we have been able to assemble to date.

Using these estimates, it would seem to us that the home environment is very significant not only because of the large amount of educational growth which has already taken place before the child enters the first grade but also because of the influence of the home during the elementary school period. Undoubtedly, the adolescent peer group must have a significant influence on the individual's attitudes toward school and learning (Coleman, 1961), but we must point out the large amount of educational growth which has taken place before the adolescent peer group occupies a prominent position in the student's life.

Since our estimates suggest that about 17% of the growth takes place between the ages 4 and 6, we would hypothesize that nursery school and kindergarten could have far reaching consequences on the child's general learning pattern. Also the approximately 17% of growth which takes place between ages 6 and 9 seems to us to suggest that the first period of elementary school (grades 1 to 3) is probably the most crucial period available to the public schools for the development of general learning patterns. We are inclined to believe that this is the most important growing period for academic achievement and that all subsequent learning in the school is affected and in large part determined by what the child has learned by the age of 9 or by the end of grade 3.

Table 4.2. Correlations between Initial Measures, Gains, and Retest Measures

| Author and Date | Tests | Period of Gain | Initial versus Retest | Initial versus Gain | Retest versus Gain |
|---|---|---|---|---|---|
| Alexander (1961) | Chicago Reading | Grades 2–8 | .66 | +.36 | .94 |
| | | 4–8 | .70 | +.13 | .80 |
| | | 6–8 | .77 | −.07 | .59 |
| | | 2–8 | .64 | +.24 | .90 |
| | | 4–8 | .76 | +.27 | .83 |
| | | 6–8 | .86 | +.01 | .52 |
| | | 2–8 | .71 | +.42 | .94 |
| Haggerty (1941) | Stanford Achievement | 6–9 | .89 | −.23 | .23 |
| Chausow (1955) | ACE Crit. Think. | 13–14 | .68 | −.28 | .23 |
| Morris (1958) | ACE Crit. Think | 13–14 | .71 | −.48 | .37 |
| Scannell (1958) | Iowa Achievement | 4–8 | .73 | −.41 | .32 |
| | | 9–12 | .91 | +.17 | .57 |
| Heston (1950) | Grad. Rec. Gen. Ed. Index | 14–16 | .90 | −.23 | .42 |
| Lannholm (1952) | Grad. Rec. Gen. Ed. Index | 14–16 | .91 | −.07 | .35 |
| Silvey (1951) | Nelson-Denney Reading | 13–14 | .83 | −.20 | .38 |
| Learned and Wood (1938) | Revised 1928 College Senior Examination | 14–16 | .89 | −.24 | .45 |
| | | 14–16 | .90 | −.23 | .52 |

NATURE OF THE GAINS

In Table 4.2 we have shown the correlations found in a number of longitudinal studies between the initial measures and the gains. In general these correlations are low, with about two-thirds of them being less than $\pm.25$. This table demonstrates the general independence of gains from initial position—a typical finding of most longitudinal research. Similar findings were pointed out in Chapters 2 and 3 on physical growth and intelligence data.

One possible explanation for such findings is that the reliability of gains scores may be so low that a great deal of error enters into measures of gain. Several writers have pointed to this phenomenon and have attempted to find ways of determining the reliability of gain scores (McNemar, 1958; Lord, 1956, 1958; Webster and Bereiter, 1961; Dressel and Mayhew, 1954). It is likely that unreliability is partially responsible for the low correlations, especially when the gains are over short periods of time. However, when long periods of growth are under consideration or when the gain measures are equal or larger than the initial measures, it is unlikely that the unreliability of the gain scores will account for the low correlations. It will be noted in Table 4.2 that the absolute magnitude of the correlations between initial measures and gains is not much greater when the gain is over four or more years than when the gain is over two years or less.

A further argument against the unreliability explanation is the level of correlations found between gain measures and final measures. It will be noted in Table 4.2 that the gain-retest correlations are much higher than the initial-gain correlations. These higher values can be explained on the basis of the part-whole relation in which the final measure includes both the initial measure and the gain measure.

Closely related to reliability is the ceiling effect of many of the tests. It is frequently found that students who are initially high on a test may make smaller gains than students who were initially low (Dressel and Mayhew, 1954; Chausow, 1955). This would explain some of the negative correlations between initial and gain scores. The point of this is that achievement tests do not have equal units at all points and it may be easier to make a large gain at the less difficult end of the scale than it is to make a smaller gain at the more difficult end of the scale. It is possible that psychometric techniques may be improved to eliminate this source of error in the tests. However, there may also be a ceiling effect in the instruction in that a teacher may be relatively satisfied with the performance of the top group of students and may make little effort to encourage further growth, whereas the same teacher may do everything possible to bring the low students up to "standard."

Another possible explanation for the low relationship between gain and initial measures is that the gains tend to be randomly distributed among students who are at various points on the initial measures. This explanation would imply that two individuals who made the same initial score might make very different gain scores. Although this does not appear to be a very satisfactory explanation since it does not account for the variations in gain, let us determine whether the results of one study show this random variation.

Alexander's (1961) study of reading comprehension at grades 2 to 8 is relevant here. In Chart 4 we have indicated the initial and final reading comprehension scores for two groups of students.* One group of students has parents in occupations requiring four or more years of higher education, while the other group of students has parents in unskilled occupations requiring less than completion of high school. If we ignore the home background of these 40 students, it would appear that the gains (indicated by the slope of the lines) are related to the initial scores in a random way.

If, however, the home background is taken into consideration, it will be noted that the students whose parents are in professional occupations have made a characteristic gain. That is, the majority of these students, regardless of initial score, makes about the same general gain from grades 2 to 8. In contrast, the students whose parents are in unskilled occupations have a much smaller gain. Here again, however, these students make about the same gain, regardless of initial score.

Thus, when two or more very different environments are involved, the gains tend to be randomly distributed. On the other hand, when a single powerful environment is involved, the gains tend to be very similar for all the students in it. Within each environment the gains are relatively independent of the individual's initial position on the measuring scale. We are inclined to the view that the low correlation between initial scores and gains is largely to be accounted for on the basis of the presence of a number of different environments which produce what appears to be random variation in the gains or on the basis of a single environment which produces very similar gains for most of the students.

The gain scores are somewhat more highly correlated with the retest

---

* This sample of 40 children was selected from a larger group of 154 children. The two sub-samples were selected to make the initial scores as comparable as possible.

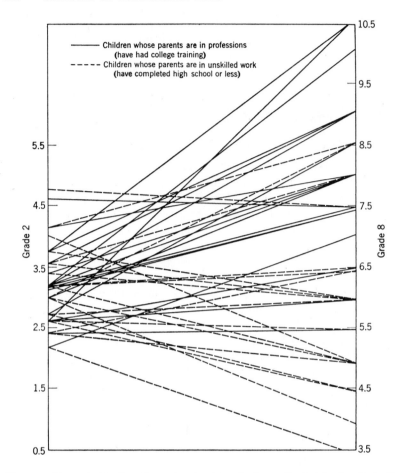

*Chart 4.   Change in Reading Comprehension from Grade 2 to Grade 8 for Pupils in Two Different Home Environments (Adapted from Alexander, 1961).*

or final measures.   The correlation between the gains and final scores is a function of the proportion of the final measure which is accounted for by the gain measure.   It is thus likely that gains over a number of years will be more highly correlated with the final measure than will the gains over a single year.

It is evident in Table 4.2 that, in general, the final measures are more highly correlated with the *initial measures* than they are with the *gain measures*.   Thus it is evident that grades based on annual measures are little more than a reflection of where the students were on the scale when they began the year.   One might question the propriety of a grading system where the grades at the end of the year could have been assigned with great accuracy before the students began the year.

Insofar as gain and final scores are concerned,

1. The correlation between gain and final measure will be very low if the gain represents a very small proportion of the final measure (on an absolute scale).

2. The correlation between gain and final measure should be near unity if all the students started at about the same point.

3. The correlation between gain and final measure will be near zero if all students make about the same gain—a condition likely to be present when all students are involved in a common set of learning experiences and are in the same powerful educational environment.

Only when the conditions suggested in item 2 are present is one justified in using final measures as the basis for assigning grades or marks.   Under the other two sets of conditions, the use of final measures as the basis for marking introduces so much error as to make this procedure highly questionable.   In these two cases, grades should be based on gains or growth rather than final status.

## EFFECTS OF ENVIRONMENTS

Most of the longitudinal studies we have summarized in Chart 3 are based on samples of students in a variety of schools and home environments.   We have demonstrated that the correlations found in these studies are very close to those expected from the Overlap Hypothesis when vocabulary development is used as an absolute scale.   That is, the correlations reflect the amount of growth which has already taken place rather than predictions of the growth which is likely to take place after the indicated year or grade.   These correlations thus represent indicators of where the group is on an academic growth curve.

We have in previous chapters pointed out that the Overlap Hypothesis represents a minimal estimate of the correlation between measures at two points in time because the measures do not take into consideration the environments under which different individuals are developing.   The measures represent collections of data about *individuals*. We have postulated that if a sample of individuals is subjected to the same environmental influences, the growth of the individuals during the period of time under consideration should be the same and that the correlation between the initial and final measure should be unity.

Some relevant evidence on this proposition is available from the research on twins and siblings reared together and reared apart. We have summarized several of these studies in Table 4.3. Here, it will be noted that the correlation between the achievement measures on identical twins reared together tend to be very high and for the most part are over +.85. In direct contrast are the much lower correlations obtained when the identical twins are reared apart. These correlations tend to be +.70 or lower. We believe the difference between the Burt (1955) results and the Newman, Freeman, and Holzinger (1937) results may be attributable to the extent to which the environments of the separated twins were different. This is supported by the correlation of +.91 reported by Newman, Freeman, and Holzinger (1937) between the differences in the educational environments of the separated twins and the differences in their scores on the Stanford Achievement Test. It is interesting to observe that the differences in achievement scores correlate only +.35 with differences in social environment and +.18 with differences in physical and health environment.

The nonidentical twins who were reared together correlate more highly than do the identical twins reared apart and almost as highly as do the identical twins reared together. Perhaps even more striking is the very high correlation between siblings reared together.*

The effect of the environment on school achievement is supported further by the correlations of about +.50 between unrelated children reared together, although there may be selective forces operating here.

In Table 4.4, we have shown the differences of separated identical twins on the Stanford Achievement Test in relation to the environmental difference ratings. For the eight identical twins with the least difference in educational environment, the average difference in their achievement scores was 6 points and the rank correlation of their scores was +.90, whereas, for the eight pairs that had the greatest differences

---

* Schoonover (1956) finds correlations of +.51 and +.49 for pairs of siblings on reading and arithmetic tests. However, these correlations are probably so different from the Burt correlations for siblings because of the difference in the variability of the samples. Burt's samples were drawn from a cross section of London schools, whereas the Schoonover samples were drawn from The Laboratory School of the University of Michigan, a relatively homogeneous group largely made up of college professors' children. Mrs. Mostovoy has completed a study of the correlation between siblings' achievement scores in which one sample is a proportionate sampling of the different social classes, whereas the other sample is restricted to a single social class. Her correlation for the heterogeneous sample of siblings on a reading comprehension test is +.85 whereas the correlation for the homogeneous sample of siblings is +.50. Thus she is able to reproduce both Burt's figures as well as Schoonover's, depending on the sample used.

*Table 4.3. Summary of the Findings of Several Studies on the Correlations between the Scholastic Achievement of Twins, Siblings, and Unrelated Children Reared Together and Reared Apart*

| | General Achievement | Reading and Spelling | | Reading | | Arithmetic | | |
|---|---|---|---|---|---|---|---|---|
| | Burt (1955) | Newman Freeman Holzinger (1937) | Burt (1955) | Husén (1959) | Wictorin† (1962) | Burt (1955) | Husén (1959) | Wictorin† (1962) |
| **Identical Twins** | | | | | | | | |
| Reared together | .90 | .96 | .94 | .89 | .95 | .86 | .87 | .83 |
| Reared apart | .68 | .51 | .65 | | | .72 | | |
| **Nonidentical Twins** | | | | | | | | |
| Reared together | .83 | .88 | .92 | .62 | .57 | .75 | .52 | .83 |
| **Siblings** | | | | | | | | |
| Reared together | .81 | | .85 | | | .77 | | |
| Reared apart | .53 | | .49 | | | .56 | | |
| **Unrelated children** | | | | | | | | |
| Reared together | .54 | | .55 | | | .48 | | |

† Based on teachers' marks at grades 2, 4, and 6. All the other correlations are based on achievement test scores.

*Table 4.4. Differences in the Achievement of Identical Twins Reared Apart as Related to Differences in the Environment*
*(Adapted from Newman, Freeman, Holzinger, 1937)*

| Differences in Years of School | Environmental Difference Ratings | | | Differences on The Stanford Achievement Tests |
| | Educational | Social | Physical Health | |
| --- | --- | --- | --- | --- |
| 14 | 37 | 25 | 22 | 69 |
| 10 | 32 | 14 | 9 | 38 |
| 4 | 28 | 31 | 11 | 35 |
| 4 | 22 | 15 | 23 | 34 |
| 5 | 19 | 13 | 36 | 14 |
| 1 | 15 | 27 | 19 | 19 |
| 0 | 15 | 15 | 15 | 0 |
| 1 | 14 | 32 | 13 | 13 |
| 1 | 12 | 15 | 12 | 16 |
| 0 | 12 | 15 | 9 | 17 |
| 1 | 11 | 26 | 23 | 6 |
| 0 | 11 | 13 | 9 | 7 |
| 1 | 10 | 15 | 16 | 19 |
| 0 | 9 | 27 | 9 | 2 |
| 2 | 9 | 7 | 8 | 2 |
| 0 | 9 | 14 | 22 | 5 |
| 0 | 8 | 12 | 14 | 1 |
| 0 | 7 | 10 | 22 | 4 |
| 0 | 7 | 14 | 10 | 8 |

in their educational environments, the average differences in their achievement scores was 28 and the rank correlation for their scores was −.09.

We may conclude from these data that the environment has a powerful effect on the educational achievement of children. We may further point out that the study of twins who are separated makes little sense unless evidence on the environments is collected. Thus, if the separated identical twins are placed in very similar educational environments, the correlation between their educational achievement would be the same as for identical twins reared together. On the other hand, if the twins are placed in extremely different environments, the correlations should approximate zero.

When we study identical twins reared apart, we make the assumption that at one time they must have been similar in the characteristics we are studying or in the *potential* for these characteristics and that any differences between the pairs must be attributable to the environmental differences. We can crudely approximate the studies of separated twins if we match unrelated children on a specific characteristic at one time and then observe the effects of different environmental conditions when we measure the pairs of children at a later point in time.

We selected the second graders from Alexander's (1961) study and found it possible to match twenty pairs of these children who had identical scores on the Chicago Reading Test but who attended different schools and had fathers at different occupational levels. For one member of each pair, the father had an occupation which required higher education, whereas for the other member of the pair the father's occupation required less than a high school education. When these children were retested on another form of the Chicago Reading Test at grade 8, the differences between the pairs averaged 2.25 grade levels. Chart 4 shows the initial and final reading scores for these two groups of students.

Put in other words, because of the way in which the samples were selected, there was a zero correlation for these pupils between reading comprehension at grade 2 and the occupational level of the fathers. However, at grade 8 the correlation between reading comprehension and the occupational level of the fathers was +.50. For these forty pupils, the correlation between their reading comprehension scores at grades 2 and 8 was +.52. The multiple correlation between reading comprehension at grade 8 and the combination of reading comprehension at grade 2 and father's occupation is +.72 (+.80 attenuated).*

This same relationship may be seen in Chart 4. It will be noted that there is very little overlap between the grade 8 reading comprehension scores of the two groups although they were equal at grade 2. Furthermore, it will be seen that within each group, the students keep much the same ranks at grades 2 and 8, although the overall ranks (when the two groups are combined) change greatly.

It will also be seen that the variability for the total group is much greater at grade 8 than the variability for each group. Thus we may conclude that much of the variability at grade 8 on reading comprehension may be attributed to the environmental differences between

---

* Of the 154 children in Alexander's study of reading comprehension there were 130 children whose father's occupation was reported. The multiple correlation between reading comprehension at grade 8 and the combination of reading comprehension at grade 2 and the father's occupation is +.79 (+.88 attenuated). In such a study all the occupations are coded and less extreme differences among the home environments are taken into consideration.

the two groups.  Here is evidence that children in extreme environments, insofar as educational stimulation is concerned, will make very different progress on a reading comprehension measure.

Much the same point can be made from the Learned and Wood study done in 1938.  There were two colleges (college A and B) in this study in which the mean scores and standard deviations at grade 14 were approximately equal (means = 527,525; standard deviations = 135,140).  When these students were retested with the same test at the end of grade 16, the mean score and standard deviation in college A were 651 and 170 whereas the retest mean score and standard deviation in college B were 542 and 166.  Thus the average student in college A gained 124 points, while the range of gains for the middle two-thirds of these students was 51 to 197 points.  In college B the average student gained only 17 points, and the range of gains for the middle two-thirds of these students was $-58$ to $+88$ points.  By the end of their college careers there was relatively little overlap in the scores or gains of these two groups of students, even though the two groups were almost identical at the end of the sophomore year of college.

In the Alexander (1961) study, the two environments represent the combined influences of home and school, whereas at the college level the direct effect of the school environment may be studied by itself.  When the two colleges in the Learned and Wood (1938) study are combined, the correlation between initial scores and gains is $+.25$ because the gains are apparently randomly distributed.  When the two colleges are separated, the correlations between the initial scores and gains for the two colleges are $+.13$ and $+.27$.  Here the correlation between initial scores and gains within each group is low because the students within each college tend to make very similar gains.

A similar phenomenon is evident in many educational studies where the effects of learning experiences are studied under "control" and "experimental" conditions.  There are many studies which could be used to illustrate the point.  Chausow's (1955) study of problem solving in a college course in Social Sciences compared a lecture versus a discussion approach to the subject.  Chart 5 shows the initial and final scores made by the initially high, middle, and low students in his two groups.  It will be noted that the gains for the students within each such group under the discussion (experimental) method is very similar.  It is as though most of the students in the discussion learning experience were affected in much the same way.  They made significant gains on the problem solving test, whereas the gains for the lecture (control) group approached a chance level.  Since the lecture

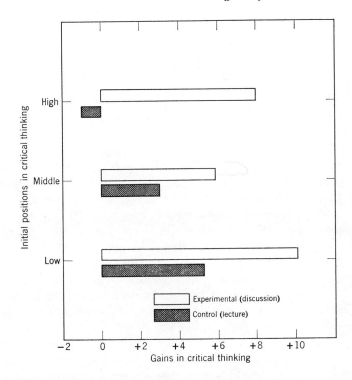

*Chart 5. Gains in Critical Thinking in Social Science Made by College Students under Discussion and Lecture Learning Experiences (Adapted from Chausow, 1955).*

method did not constitute a powerful and very similar environment for all, the gains for the three subgroups under this method were more variable. The effect of the discussion method was to produce a very different type of growth in problem solving than was produced by the lecture method. The point of this illustration is that a powerful set of learning experiences tend to alter individuals in very similar ways, and under appropriate conditions the amount of gain will be almost the same for all individuals in spite of the differences in initial measurements.

It will be evident from the foregoing discussion that we are using the term "educational environment" to encompass both the home and out-of-school environment as well as the school environment. It is likely that the home environment will be more powerful than the typical school environment in the early years of school, whereas the school environment will be more powerful than the home environment during the college years, especially when students are living away from their homes.

We hesitate to use the terms, but we do believe the educational environments may vary from "deprived" to "abundant" or stimulating. We regard the two colleges in the Learned and Wood (1938) study as illustrating these two poles. The effect of the college A environment was to increase the students' score on a general achievement test by an average of 124 points, whereas in the college B environment the average gain was 17 points. Although we do not know anything about these two colleges since the authors did not reveal the identity of the colleges nor did they provide descriptive information about these two institutions, we infer the deprivation and the stimulation from the changes in test scores.

In the Chausow (1955) study, the two groups of students came from similar home backgrounds and had the same initial performance. Here two very different types of learning experiences constituted the extreme environments. It should be noted that the term environment is not entirely appropriate in describing Chausow's experimental learning experiences since these students were taking three or four other courses at the same time and the learning experiences in these other courses were patterned very much in the same way as the lecture method for the control set of learning experiences. Under such conditions, one would expect a smaller effect from a special set of learning experiences (that is, the discussion method) since they are in some conflict with the more traditional learning experiences that the students are simultaneously undergoing. We would expect the impact of an environment to be greater as it is more pervasive and as it sets up a highly integrated pattern of experiences and reinforcements.

Dressel and Mayhew (1954) studied the gains of college freshman on a problem solving test. They found almost no gains on this test for students in colleges which placed no emphasis on this ability in the freshman courses. When the college had one course which gave some emphasis on problem solving, but the other courses did not, the gains were small but significant. In several colleges, problem solving was the major objective of all the courses taken during the freshman year and most of the learning of the students throughout the year emphasized this type of learning. In these colleges, Dressel and Mayhew found massive changes on this test of problem solving from the beginning to the end of the year. It seems clear that as far as problem solving was concerned the third group of colleges represents powerful, stimulating, and abundant environments, whereas the first group of colleges more nearly represents the "deprived" type of environment.

In the Alexander (1961) study (see Chart 4) we believe we have a clearer case of a deprived and an abundant educational environment in which both home and school combine to produce quite different levels of educational growth.

In terms of their effects on the students we may refer to one as a "deprived educational environment," whereas the other may be referred to as an "abundant or stimulating educational environment." What is the difference between such environments, not only in their effects but also in the process by which they bring about these effects? Such environments encompass not only what happens within the schools but also in the home as well as out of school.

Although we cannot fully describe the difference between a deprived and a stimulating educational environment,* we believe the differences in the Alexander (1961) study to center around four major points.

1. The value placed on school learning by parents and students. Symptomatic of this is the amount of education the parents have had (beyond high school for the average parent in school A and less than tenth grade for the average parent in school B). Especially symptomatic is the extent to which the students plan for higher education (80% in school A as compared with about 25% for school B).

2. The reinforcement of school learning by the home. Symptomatic of this is the interest taken in the affairs of the school as demonstrated by the PTA and parents' visits to determine what is going on in the school. These differ enormously for the two schools, with a large percent of the parents in school A being members of the PTA as compared with very few of the parents in school B. The adjustment teacher reports that only a rare parent has visited school B, whereas the limiting of parent visits is a pressing problem in school A.

3. Economic returns from education. The students in school A see their future economic welfare being determined by adequate performance in school, whereas the students in school B have difficulty in seeing a clear relationship between the jobs they will secure later and their performance in the school.

4. Morale and training of the school staff. The teachers look on a position in school A as a highly desirable one, whereas they do not see great merit in holding a position in school B. Symptomatic of this is the fact that in school A the teachers are more experienced and other teachers desire to transfer into this school. Many teachers seek to transfer out of school B.

* School A is the abundant environment and school B is the deprived environment. These are combinations of home and school environments.

A somewhat different approach to the study of environments related to educational achievement was undertaken by Dave (1963). He hypothesized, on the basis of the literature, that the home environment relevant to educational achievement might be studied in terms of six variables.

1. Achievement press
2. Language models in the home
3. Academic guidance provided in the home
4. The stimulation provided in the home to explore various aspects of the larger environment
5. The intellectual interests and activity in the home
6. The work habits emphasized in the home

These six variables were further broken down into 22 process characteristics which were used to summarize and rate the mothers' responses to an interview schedule. Sixty mothers were interviewed and the ratings on their responses were related to the scores of their children on a battery of achievement tests taken at the end of the fourth grade of school.

The overall index of the home environment had a correlation of +.80 with the total score on the entire achievement battery.

Dave has analyzed the relations between particular aspects of the home and the scores on the different parts of the achievement battery. In general, the correlations are highest with the tests of word knowledge and reading and they are lowest with spelling and arithmetic computation. This suggests that the home has the greatest influence on the language development of the child and the least influence on skills primarily taught in the school.

Dave's correlation of +.80 may be contrasted with the much lower correlations (usually less than +.50) between school achievement and other indices of the environment such as socio-economic status, education of parents, occupational status, or social class. This approach to the study of the interaction between the home and the child has much promise. If supported by further research, these techniques may enable the school to analyze the home environment and to determine the best strategy for the school and the home to provide the environmental conditions necessary for school achievement.

This approach also makes it clear that parents with relatively low levels of education or occupational status can provide very stimulating home environments for educational achievement. Dave's research demonstrates that it is what the parents *do* in the home rather than their *status* characteristics which are the powerful determiners in the home environment. The relationship between the interactive processes and the parent's status is relatively low.

## SPECIFIC TYPES OF LEARNING

Throughout this chapter we have been concerned with reading comprehension and overall indices of educational achievement. The absolute scale on vocabulary development does, in general, fit the correlational data reported in longitudinal studies of these rather general learning patterns.

However, it seems clear that all types of learning are not describable by a single curve of development. Gardner (1947) has developed a technique for constructing equal-interval scales by the use of normative data on achievement tests. This technique, referred to as the K scale, uses the mean and standard deviation at one age or grade as the basis for scaling the scores at other age or grade levels. Gardner has applied this technique to the data available on the Stanford Achievement Tests (Kelly, Gardner, et al., 1953) and we have plotted the developmental scale for four of the subtests of this achievement battery (see Chart 6). Although Gardner has constructed equal-interval scales, he does not have an absolute scale with a zero point.

It is evident in Chart 6 that the language test shows rapid development in the first seven grades and then comes to a virtual plateau, whereas the arithmetic computation test shows little development until about grade 5, with continued rapid development from grades 5 to 13. Science shows relatively rapid development from grades 1 to 5, relatively slow development from grades 5 to 11, and then again rapid development from grades 11 to 13. The social studies tests shows continued rapid development from grades 1 to 13.

It is evident that the developmental curves of achievement in specific subjects may differ somewhat from each other as well as from the more general absolute scale we have derived on the basis of vocabulary development. Further research is needed on the techniques of absolute scaling as well as their application to a variety of data on academic achievement growth. If these scales are appropriately developed, they should yield theoretical values which may be compared with the observed values found in longitudinal studies.

Both the theoretical and observed values should be of great importance in describing the effects of present curricula in the schools and they should be a major source of data in planning new curriculum development and educational research.

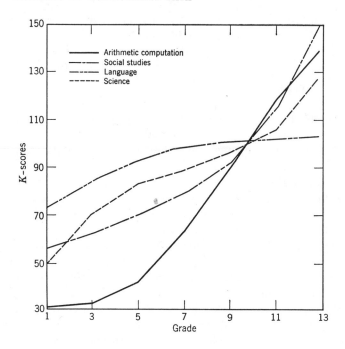

*Chart* 6.   *Growth of Academic Achievement in Selected Areas (Adapted from T. L. Kelley,*
*E. F. Gardner et al.,* 1953).

SUMMARY AND IMPLICATIONS

There have been very few long-term longitudinal studies of educational
achievement in spite of the fact that existing school records represent
excellent sources of data for such studies.   When the longitudinal
studies that have been done on teachers' marks, general achievement
as represented by the total score on an achievement test battery, and
reading comprehension are compared with each other, the patterns of
relationships with a criterion point vary greatly.   However, when the
different measures are corrected for the unreliability of the instruments, the patterns of relationships become very similar.   Although
we have not made the computations because of the lack of comparable
norms on the different instruments, we believe the pattern of relationships would become even more similar if the correlations were all corrected to a common variability.

The patterns of relationships of the three indices we have used
closely approximate the values inferred from an absolute scale based
on normative data on vocabulary development.   This absolute scale

yields theoretical estimates which rather closely fit longitudinal achievement data gathered over the past thirty years at the elementary, secondary, and college levels.  Perhaps the most important implication of this is that we can use theoretical values in education and educational research and that deviations of observations from these theoretical values can be explained by appropriate analyses. For example, from the absolute scale of vocabulary development we estimated the theoretical correlation between achievement indices (grades, general achievement as measured by a test, etc.) at the 12th versus 13th year of school to be about $+.95$.  However, the correlations between high school grades and college freshmen grades are usually of the order of $+.50$.  The search for the explanation of the difference between the two values led us to question the equivalence of grades from different schools or colleges and to the development of techniques for equating grades from different educational institutions. The corrected or scaled grades at the high school and college level correlate about $+.80$ (Bloom and Peters, 1961).  However, the difference between $+.80$ and $+.95$ still needed to be explained.  We found most of the difference in these two values to be accounted for by the unreliability of the high school and college grades.  Highly reliable achievement test batteries administered to students before and after one or two years of college reveal correlations of approximately $+.92$, which is very close to the theoretical value derived from the absolute scale.

Although more precise work is needed to secure better absolute scales and more accurate theoretical values than are at present available, we believe that education will profit greatly if its research problems arise from the need to explain differences between theory and observation and differences between theoretical and observed values.

The absolute scale of vocabulary development and the longitudinal studies of educational achievement indicate that approximately 50% of general achievement at grade 12 (age 18) has been reached by the end of grade 3 (age 9).  This suggests the great importance of the first few years of school as well as the preschool period in the developing of learning patterns and general achievement.  These are the years in which general learning patterns develop most rapidly, and failure to develop appropriate achievement and learning in these years is likely to lead to continued failure or near failure throughout the remainder of the individual's school career.  The implications for more powerful and effective school environments in the primary school grades are obvious.  On the other hand, this research raises serious questions about the value of educational remedial measures at later stages.

We have also analyzed some of the effects of the school and home environment on learning. It is evident that when the school and home environments are mutually reinforcing, learning is likely to be greatest. Furthermore, powerful and consistent school environments are likely to have the greatest effect on particular kinds of learning. The nature of the learning environment is most critical during the periods of most rapid change in learning—the early years of school. It is also likely that the greatest changes may take place in the individual when he enters a new level of school environment, that is, high school or college, if the new environment is different from the previous one and if it is a powerful and consistent learning environment.

The analysis of the relationships between gains and initial and final status raises serious questions about grading practices which are based on status measures rather than gain or growth measures. The discrepancies between grades based on status measures and grades based on gain measures are likely to be greatest for those educational achievement characteristics which are most stable. Under present grading practices for stable characteristics individuals are likely to be repeatedly rewarded or punished for characteristics they possess at the beginning of a school term, whereas the gains they make during the school term are largely ignored. Thus two individuals may make equal progress during a given school year but be given very different grades at the end of the year because they started at different points at the beginning of the year. The consequences of repeated success or failure over several school years must surely have major effects on the individual's view of himself and his attitudes toward school and school learning.

In this chapter we have made use of vocabulary development as the basis for an absolute scale. This appears to be an excellent scale which fits the longitudinal data on teachers' marks, overall scores on achievement test batteries, and reading comprehension. However, it is unlikely that this absolute scale will be appropriate for more specific achievement in the school subjects. Gardner (1947) has devised an equal-interval scaling technique which he has applied to normative data based on the Stanford Achievement Tests. These more specific scales (Kelley, Gardner, et al., 1953) suggest that the most rapid period of growth in arithmetic computation takes place during grades 5 to 13, the most rapid period of growth in science takes place during grades 7 to 13, whereas the most rapid period of growth in social studies takes place from about grades 9 to 13. Further research is needed to develop precise absolute scales. We also need to know whether the present curves of development are reflections of the way in which the school curricula are now organized or whether they are reflections of more basic growth and development in the individual student. In either case, curriculum development and the setting of educational objectives must take into consideration the points in the individual's development when educational growth in a particular type of learning is most effectively influenced by learning experiences in the school and home environments.

REFERENCES

Aaron, S., 1946. The productive value of cumulative test results. Ph.D. Dissertation, Stanford Univ.

Adam, C. L., 1940. The stability of achievement differentials of a high school student. *J. Exp. Ed.*, **9**, 64–86.

Alexander, M., 1961. Relation of environment to intelligence and achievement: a longitudinal study. Unpublished Master's Study, Univ. of Chicago.

Bloom, B. S., and Peters, F., 1961. The use of academic prediction scales for counselling and selecting college entrants. Glencoe, Illinois: Free Press.

Bryan, F. E., 1953. How large are children's vocabularies? *Elem. School J.*, **54**, 210–16.

Burt, C., 1955. The evidence for the concept of intelligence. *Brit. J. Ed.*, **25**, 158–177.

Byrns, R., and Henmon, V. A. C., 1935. Long range prediction of college achievement. *Sch. and Soc.*, **41**, 877–880.

Chausow, H. M., 1955. The origanization of learning experiences to achieve more effectively the objective of critical thinking in the general Social Sciences courses at the junior college level. Unpublished Ph.D. Dissertation, Univ. of Chicago.

Coleman, J. S., 1961. The adolescent society. Glencoe, Illinois: The Free Press.

Coleman, W., and Cureton, E. E., 1954. Intelligence and achievement: the "Jangle Fallacy" again. *Ed. and Psych. Meas.*, **14**, 347–51.

Dave, R. H., 1963. The identification and measurement of environmental process variables that are related to educational achievement. Unpublished Ph.D. Dissertation, Univ. of Chicago.

Dressel, P. L., and Mayhew, L. B., 1954. General education: Explorations in evaluation. Washington: American Council on Education.

Ebert, E., and Simmons, K., 1943. The Brush Foundation study of child growth and development. I. Psychometric tests. *Monogr. Soc. Res. Child. Developm.*, **9**, No. 2, 1–113.

Finch, F. H., and Nemzeh, C. L., 1934. Prediction of college achievement from data collected during the secondary school period. *J. Appl. Psych.*, **18**, 454–460.

Gardner, E. F., 1947. The determination of units of measurement which are consistent with inter and intra grade differences in ability. Unpublished Ed. D. Dissertation, Harvard Univ.

Haggerty, L. H., 1941. An empirical evaluation of the accomplishment quotient: A four year study at the junior high school level. *J. Exp. Ed.*, **10**, 78–90.

Hartmann, G. W., and Barrick, F., 1934. Fluctuations in general cultural information among undergraduates. *J. Ed. Res.*, **28**, 255–264.

Heston, J. C., 1950. Educational growth as shown by retests on The Graduate Record Examination. *Ed. and Psych. Meas.*, **10**, 367–370.

Hicklin, W. J., 1962.  A study of long range techniques for predicting patterns of scholastic behavior.  Unpublished Ph.D. Dissertation, Univ. of Chicago.

Hildredth, Gertrude, 1936.  Results of repeated measurement of pupil achievement. *J. Ed. Psych.*, **21**, 286–296.

Hiremath, N. R., 1962.  Relation of teacher's grades to environment: a longitudinal study.  Unpublished Master's Dissertation, Univ. of Chicago.

Husén, T., 1959.  Psychological twin research, *Univ. of Stockholm Studies in Educational Psychology*, No. 3.

Kelley, T. L., 1914.  Educational guidance.  Teachers College, Columbia University, New York.

Kelley, T. L., 1927.  Interpretation of educational measurement.  Yonkers, New York: World Book Co.

Kelley, T. L., Gardner, E. F., et al., 1953.  Directions for administering the Stanford Achievement Tests.  Yonkers, New York: World Book Co.

Krantz, L. L., 1957.  The relationship of reading abilities and basic skills of the elementary school to success in the interpretation of the content materials in the high school.  *J. Exp. Ed.*, **26**, 97–114.

Kvaraceus, W. C., and Lanigan, M. A., 1948.  Pupil performance on the Iowa Every Pupil Tests of Basic Skills administered at half-year intervals in the junior high school.  *Ed. and Psych. Meas.*, **8**, 93–100.

Lannholm, G. V., 1952.  Educational growth during the second two years of college.  *Ed. and Psych. Meas.*, **12**, 645–653.

Learned, W. S., and Wood, B. D., 1938.  The student and his knowledge.  New York: The Carnegie Foundation for the Advancement of Teaching.

Lord, F. M., 1956.  The measurement of growth.  *Ed. and Psych. Meas.*, **16**, 421–437.

Lord, F. M., 1958.  Further problems in the measurement of growth.  *Ed. and Psych. Meas.*, **18**, 437–451.

Lorge, I., 1934.  Retests after ten years.  *J. Ed. Psychol.*, **25**, 136–141.

McNemar, Q., 1958.  On growth measurements.  *Ed. and Psych. Meas.*, **18**, 47–55.

Morris, J. B., 1953.  Critical thinking gains and achievement in a Social Science course.  Unpublished Ph..D. Dissertation, Syracuse Univ., New York.

Mostovoy, J. L., Sibling resemblance in intelligence and achievement as related to parental education.  Master's Thesis in progress, Univ. of Chicago.

Newman, H. H., Freeman, F. N., and Holzinger, K. J., 1937.  Twins: a study of heredity and environment.  Chicago: Univ. of Chicago Press.

Scannell, D. P., 1958.  Differential prediction of academic success, from achievement test scores.  Unpublished Ph.D. Dissertation, State Univ. of Iowa.

Schoonover, Sarah, 1956.  A longitudinal study of sibling resemblances in intelligence and achievement.  *J. Ed. Psych.*, **47**, 436–442.

Seashore, H. G., and Eckerson, L. D., 1940.  Measurement of individual differences in general English vocabularies.  *J. Ed. Psychol.*, **31**, 14–38.

Silvey, H. M., 1951.  Changes in test scores after two years in college.  *Ed. and Psych. Meas.*, **11**, 494–502.

Smith, M. K., 1941.  Measurements of the size of general English vocabulary through the elementary grades and high school.  *Genetic. Psychol. Monogr.*, **XXIV**, No. 2, 313–45.

Spaulding, G., 1960.  Another look at the prediction of college scores.  Educational Records Bureau.  Unpublished report for the College Entrance Examination Board.

Townsend, Agatha, 1944.   Some aspects of testing in the primary grades.   *Ed. Records Bulletin*, **40**, 51–54.

Townsend, Agatha, 1951.   Growth of independent school pupils in achievement on the Stanford Achievement Test.   *Ed. Records Bulletin*, **56**, 61–71.

Traxler, A. E., 1941.   A study of the revised edition of the Stanford Achievement Test. *Ed. Records Bulletin*, **35**, 51–57.

Traxler, A. E., 1950.   Reading growth of secondary school pupils during a five year period.   Achievement Test Program in Independent Schools and Supplement Studies.   *ERB Bulletin*, **54**.

Webster, H., and Bereiter, C., 1961.   Reliability of difference scores for the general case.   Unpublished study.

Wictorin, M., 1962.   Bidrig till räknefärdighetens psykologi.   En tvillingundersökning. Goteborg. Ph., D. Thesis.

*Chapter Five*

# INTERESTS, ATTITUDES, AND PERSONALITY

INTRODUCTION

The stability of general intelligence, specific aptitudes, and scholastic achievement is an empirical finding which has theoretical as well as practical significance. However, one does not find much in the literature in the way of a theory on the development of these characteristics. To be sure, there have been genetic and environmentalistic viewpoints about such development, but it is difficult to find a comprehensive theory* or model which could give direction to the research on the development of these characteristics. One thus has observations and evidence of early development and stabilization of particular characteristics in advance of a general theory of how these *characteristics* are developed and altered.

In the general area of personality, the situation is quite the reverse. A great many theories and models of personality development have been formulated, most of which include some statements about the characteristic development at selected stages in the growth process. Such theories, at least qualitatively, describe the course of personality development and suggest the way in which this development may be altered or affected by various experiences. Most of the theories are stated in qualitative and dynamic form, and quantitative features of the development can *only* be inferred from the general form of these statements. What is common to most of these theoretical formulations is the assumption of very rapid personality development in

---

* We recognize that many assertions have been made about the formation of these characteristics, but these have rarely been cast in the form of a major theory or model.

the early years of infancy and childhood, the possibility of marked changes in the adolescent period, and the likelihood of small change during adulthood and maturity.

Sanford (1962, p. 259) has plotted a graph to show his interpretation of the quantitative changes in ego development and impulse control over the total life span. This is one of the few attempts at quantitative expression of personality development. We shall consider Sanford's graph later in this chapter.

However, the major point to be made here is that the theoretical formulations of personality development are far more complete than are the attempts at theoretical formulations in other areas. What is missing is the attempt to relate quantitative changes in personality development to these theoretical formulations. Thus, although in other chapters, we have some excellent longitudinal data which reveal very clear and consistent patterns of growth and development with theoretical formulations *yet* to be constructed, it is in the personality area that the theoretical formulations have long preceded attempts to organize the quantitative data on the stability or change of personality characteristics.

In this chapter we shall attempt to summarize the longitudinal data which bear most directly on the emotional or affective components of the individual's development. For our purposes, this will include measures of interests; measures of opinions, attitudes, and values; and measures of other personality characteristics. We do not believe these measures are purely measures of affective characteristics since many of them include cognitive components as well. What does bring some unity into this chapter is that the measures do attempt to get at ways in which the individual is seen by himself or others; ways in which the individual reacts to others; and the ideas, views, and aspects of the world which the individual regards as important or desirable. These measures do not have a right and wrong polarity; rather they describe how individuals behave and react and how they view objects, ideas, people, and self.

This has been a difficult area to measure and one is struck by the lack of a unified view about how to measure these characteristics and the general lack of instruments which are regarded as clearly valid and useful by the workers in the field. The Stanford-Binet is viewed by workers in the field as the major instrument for determining general intelligence. Its long history, the many studies which have been based on it, and the great respect for Terman's work make this instrument stand out as a virtual criterion measure against which other instruments may be validated. We, as yet, do not have instruments in the affective area which loom so large and which give us a criterion against which to compare other instruments or findings.

Furthermore, we do not have a large number of longitudinal studies making use of the same or similar instruments. We must piece together as best we can the studies and measures which are available. Thus, although we do believe the results in the preceding chapters give a definitive picture of stability and change for selected characteristics of physical growth, intelligence and aptitudes, and school learning, we regard this chapter as yielding only the broad outlines of stability and change for interests, attitudes, and personality. The definitive studies are yet to be made on these complex processes.

In the other chapters, we have used measures at ages 18 to 20 as criterion measures because most of the characteristics were studied intensively during the period from birth to ages 18 or 20, and because many of the selected characteristics reached approximately full development by these criterion ages. In this chapter we will bring the longitudinal data together for various segments of the life span. For some characteristics we will have a great deal of evidence in the period from age 18 onward, whereas for other characteristics the major data will be in the early childhood to adolescent stage. It is to be hoped that putting these different segments in relation to each other will help us to get a glimpse of the general course of development of these characteristics. Putting these segments in relation to the theoretical estimates, to be discussed in the following section, may be of help in describing the developmental features of these characteristics.


THEORETICAL ESTIMATES

Throughout this work we have made use of the Overlap Hypothesis as a basis for organizing the various longitudinal findings. In each chapter we attempted to find a developmental scale which could be used as a basis for theoretical estimates of the relationships between measurements made at selected ages. For the most part, these theoretical estimates, when adjusted for the unreliability of the measures and the variability of the samples, yield good approximations to the observed values.

We have searched carefully for theoretical values which might have relevance for the development of interests, attitudes, and other personality characteristics. The problem is one of finding developmental scales which are relevant to change and development in personality and other affective characteristics. The literature on personality is amply supplied with qualitative material on the characteristic stages in personality development and the effects of particular difficulties at each stage.    However, there are no quantitative estimates of development in this area which would approximate the absolute scale for the development of height, the absolute scales of Thorndike and Thurstone for the development of intelligence, or the vocabulary scale we have used to estimate the development of school achievement.

Sanford (1962) has suggested curves to represent the development of ego maturity and impulse control.    In constructing these curves, Sanford indicated that he had inferred the shape of the curves from the psychoanalytic literature and in particular had made use of the writings of Anna Freud.    It is clear that Sanford did not intend these curves to be more than a schematic representation of personality development.

In Chart 1 we have reproduced Sanford's curves.    We have inserted our own age estimates for Sanford's developmental periods and we have inserted a percentage scale which is little more than an equal division of the total amount of growth depicted by each curve.

Using the crude approximations we have made on Sanford's curves, it is suggested that ego development reaches a virtual plateau at about age 25, whereas about 40 % of mature development is reached by about age 7.    On the basis of this scale, we might expect about 80 % of ego development to be attained by age 18.    Using different ages and differ-

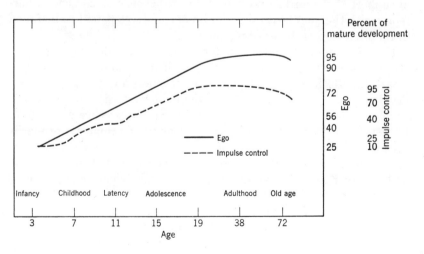

*Chart 1.    Scales for the Development of Ego and Impulse Control (Adapted from Sanford, 1962).    [Values inserted by the author.]*

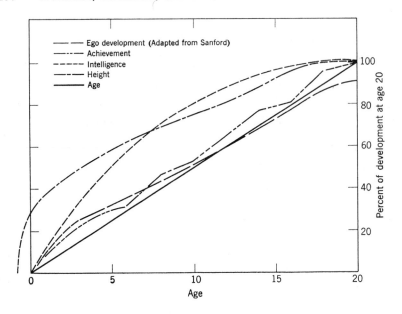

*Chart 2.   Curves of Development for Selected Characteristics.*

ent percentages would make some difference in this picture of ego development.    Although we are unable to find other quantitative data in support of Sanford's curve, we believe this curve is in general harmony with the qualitative description of ego development to be found in the literature of dynamic psychology.

It is of interest to contrast Sanford's curve of ego development with the curves of development for height, intelligence, and vocabulary (see Chart 2).    Both height and general intelligence show more rapid early development and earlier plateaus than do Sanford's curve for ego development.    We have hesitated to use Sanford's curve as the basis for securing theoretical estimates of personality development primarily because we cannot relate each of the characteristics, for which longitudinal data are reported, to ego development.    Furthermore, we can find no other quantitative developmental scale which might be used as a check on Sanford's curve.

What seemed most simple for comparative purposes is an age curve of development (see Chart 2).    If a characteristic develops in equal amounts per unit of time, it would approximate an age curve of development.    Although we have found that most stable characteristics do *not* develop in equal amounts per unit of time, we can use an age curve of development as one basis for determining whether a particular characteristic does indeed develop differentially over time. The contrast between the quantitative features of each characteristic's development and the age curve of development does offer one way of highlighting the developmental features of selected characteristics.

It will be seen in Chart 2 that all the other curves show much more rapid development in the first 5 or 6 years of life than would be expected on the basis of an age curve. Thus Sanford's curve suggests about one-third of the ego development at age 20 is likely to be attained by age 5, whereas an age curve would suggest that only a fourth should be attained by age 5. Sanford's curve suggests the beginning of a plateau of ego development at about age 18, whereas an age curve would have no plateau. Although we believe Sanford's curve is probably generally appropriate for ego development and related characteristics, we will make use of the age curve because it requires the fewest assumptions. It is to be noted that *over* short intervals in the period ages 7 to 15, the *slopes* of many of the curves in Chart 2 are such that approximately the same estimates would be derived from each curve. For longer intervals, for the early years (before age 7), and for the later years (after age 15), the results would be quite different for the different curves.

In the studies of height growth, we have very clear notions of the absolute amount of height developed by each age and by maturity. We have used the same measurement technique (a meter scale or yardstick) at each age and we are quite certain that we are measuring the same characteristic at each age. We are less certain of our scale and of the comparability of the measurements at various ages in the studies on intelligence and school achievement, but we are able to secure reasonable approximations to an absolute scale and to comparability of measurements. With such data we are able to draw inferences about the amount of growth or development by each age period in relation to the growth achieved by maturity.

In contrast, in the personality area we have no scale for determining the amount reached by maturity. We are not even certain as to what an absolute scale of aggression, dependence, etc., would mean at a particular age. Nor are we certain that we are measuring comparable characteristics at each age. What do we mean by aggressiveness in children and how may this be compared to aggressiveness in adults? Faced with problems like this we must reject any notions of interpreting personality changes in terms of an *absolute* scale of development.

What we are able to do is compare individuals on a *relative* scale at one age and to relate these results to those obtained on another relative scale at a different age. Thus, is John at age 3 relatively high on aggressiveness as compared with other 3-year-olds? Also, is John at age 14 relatively high on aggressiveness as compared with other 14-year-olds? We can then ask the question of how *consistent* are the ratings or measurements at the two ages. We can determine whether this consistency is greater than chance and we can also ask whether it is greater than we might expect on the basis of an age curve of development. This means that Anderson's Overlap Hypothesis cannot be directly tested with personality characteristics, since it is based on the idea of absolute development and comparable scales. We can, however, make use of the more general Overlap Hypothesis in which we shall determine the proportion of the variance at one age which is accounted for by an earlier age $(r^2_{X_1 X_2})$. Thus we can determine the level of *consistency* in the relative ratings at two different ages, even though we would hesitate to draw an absolute developmental curve for the particular personality characteristic. The level of consistency then will be interpreted in terms of the level of stability of a characteristic, although stability cannot be expressed in terms of absolute development. It is with these reservations that we can attempt to draw a picture of the stability and change of selected personality characteristics.

## ORGANIZATION OF THE LONGITUDINAL FINDINGS

It is difficult to find many longitudinal studies that have made use of the same instruments. Consequently, we have few studies which can be directly compared with each other to determine the consistency of results on a single instrument with different samples of subjects, at different times, and with different research workers. Furthermore, there is a dearth of longitudinal studies which follow the same sample with the same instrument administered at regular time intervals. Because of these limitations, we have had to piece out a picture of the development of each characteristic by putting the results of several studies in some relation to each other.

Another difficulty we have experienced with interests, attitudes, and personality characteristics is the different types of techniques used to gather the evidence. These vary from self-report techniques, ratings, and observational techniques to more carefully controlled experimental measurement techniques. Some of these techniques may be viewed as measuring conscious and public responses and behaviors, whereas others attempt to get at more private thoughts and feelings or at even more deep-seated and relatively unconscious thoughts and behaviors. We regard these as very different aspects of the person and there is evidence (Getzels and Jackson, 1963) that there is relatively little relationship among the measurements of characteristics at the different levels.

In order to structure our analysis of the available longitudinal data, we have divided the measurements into those which deal with interests, attitudes and values, and personality characteristics. We have further subdivided each of these types of evidence into three groupings.

1. The characteristics of individuals as seen by others. This includes observations of the individual by a research worker (psychologist, social worker, etc.); observations made by less competent and trained observers such as parents, peers, and others; as well as rating and judgments made by teachers, supervisors, etc. Most of these observations are summarized in a judgmental form such as a rating or some other quantitative classification of the individual's behavior, in a descriptive classification which may describe or classify the characteristic without necessarily placing it on a scale, or by some more general rating which describes how the individual observer feels about or views another person or his behavior, such as a sociometric rating. This last group of ratings may reveal as much about the observer as about the person or characteristic being observed.

2. Another group of techniques attempts to secure measurements in such a way as to minimize the amount of conscious control the subject may have about his responses. These may include the Rorschach Test, Thematic Apperception Test, drawing tests, etc.; or, they may include cognitive-perceptual tasks in which the subject is placed in an experimental-measurement situation which he is expected to see as a problem solving task. This group of techniques also includes records of behavior under normal conditions which are then translated into measurements, for example, the number of books read in a period of time, the choices the individual makes of the activities he participates in, or an essay on some topic which is then analyzed for a different purpose than the subject had in mind (for example, an English theme which is analyzed for evidence of particular personality characteristics).

3. The third group of techniques is the self-report the individual makes on a questionnaire, check list, or related procedure. The subject is asked to respond to a number of questions as accurately as he can. For the most part, the subject is in control of his responses and he may deliberately (or unconsciously) bias them in some way. Thus the self-report provides us with the responses of the subject, but the

meaning to be given to the responses is not always clear.   The tester is faced with the task of making whatever sense he can out of the responses made by the subject.   In addition to the questionnaire, self-reports may also include biographical statements, statements of purpose and plans by the subject, or responses in structured and unstructured interview situations.

## CHARACTERISTICS OBSERVED BY OTHERS

The judgments made by persons who are in a position to observe an individual over time or under specified conditions represent one technique for quantifying personality characteristics.   These judgments, in the form of ratings, are especially useful when applied to a specific characteristic either observed directly or inferred from case studies of the subject.   Such ratings, when made by competent observers, are likely to be useful indicators of those personality characteristics which are reflected in overt behavior and in interpersonal relations.   It is likely that more deep-seated emotional characteristics can be detected only by the most expert and sensitive observers.

Undoubtedly, differences in the competence of observers will account for some of the differences in the ratings of subjects when two or more observers are used.   The degree to which the behaviors are overt and observable will also account for variations in the agreement of observers.   Insofar as possible, we will restrict our attention to those characteristics where the observers' ratings correlate with each other at the level of +.80 or higher.   To attempt to consider personality characteristics with lower levels of observer reliability would introduce so much error variance as to make our data of questionable value.

In the following section, we shall select a number of personality characteristics from longitudinal studies and show the degree of stability observed as contrasted with the estimated values inferred from an age curve of development.   Where possible, we shall show the results for males and females separately.   Where several studies have obtained ratings on a particular characteristic or on closely related characteristics, we shall present the results of the different studies and discuss the results in relation to each other.   However, although characteristics may be given similar or closely related names, it is quite possible that very different behavioral manifestations were studied by the different workers.

We shall also include the results of a number of sociometric studies under this general classification of observational measures.   Although sociometric ratings may reveal as much about the rater as they do about the individual rated, the agreement of a number of the raters about the popularity or social desirability of an individual is likely to yield an excellent measure of the way in which an individual affects others.

For the ratings on selected personality characteristics, we shall draw most heavily on the results of four studies.

1. Kagan and Moss (1962) had research workers observe individuals in the Fels study from ages 0 to 3 to adulthood (19 to 29). The case studies, with these observations, were then made available to raters who rated the individuals on selected characteristics using rating scales devised for this purpose. The ratings were made so as to secure independent ratings for each set of observations and to maximize the independence of the ratings on each individual. These ratings are somewhat better for our purposes than the Macfarlane results (see item 4) because they are based on primary data about the individual as observed by a trained worker. On the other hand, the rather limited amount of observation made by the worker on each individual at each age period does provide some limits on the value of the data.

2. Peck and Havighurst (1960) followed a group of children from ages 10 to 16. The basic data were obtained by a psychologist who observed the subjects at these age periods. The same psychologist made the ratings. This study is likely to show higher levels of stability than the other studies because the same person did both the observations and the ratings and thus may, in part, recall previous observations and ratings.

3. Tuddenham (1959) followed a group of individuals in the California Growth Study from ages 16 to 36. Psychologists observed the individuals at age 16 for relatively brief periods of time in several situations and then rated them on a large number of characteristics. This was repeated under somewhat different conditions at age 36. While a large number of the characteristics had interrater reliabilities of less than +.80 (probably because of the limited amount of data available in the brief observations), we have selected only those characteristics which had the highest interrater reliabilities.

4. Macfarlane (1954) had social workers interview mothers of the children in the Child Guidance Study from ages 5 to 14. These interview data were then made available to two raters who rated the children on selected characteristics using rating scales devised by the research workers in this study. Efforts were made to have the ratings for each age and for each characteristic made independently of the

other ratings on the same child.    It is likely that these data are somewhat suspect since the mother (rather than the child) was the source
of the basic data and the stability or lack of it may be in the mother's
observations of the child and in her reports of what she observed.

### Sociometric Ratings

Perhaps the most precise quantitative estimates of personality characteristics as observed by others is to be found in sociometric ratings.
This technique, originally developed by Moreno (1934), systematically
records the ratings each member of a group makes about other numbers of the group.    The large number of ratings on each person and
the possibility of having each rating on a particular highly focused
characteristic make it possible to secure as reliable measurements as
are obtained in excellent psychometric task-oriented instruments.
Thus Thompson and Powell (1951) report test-retest reliabilities from
.85 to .92, while Grossman and Wrighter (1948) report Spearman-
Brown reliabilities of .93 to .97.

It is unfortunate that we have not been able to find longitudinal
studies of sociometric measurements over more than a four year
interval.    However, one of the most revealing studies is by Cannon
(1958) who obtained sociometric data on a group of students in a
rural high school in Nebraska.    This was a community in which very
few new students entered the school, so that the composition of the
group, except for dropouts, was not markedly altered during the four
years.    In Chart 3 we have plotted the correlations (corrected for

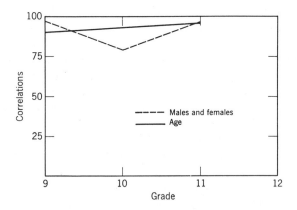

*Chart* 3.    *Correlations (Attenuated) between Sociometric Position at Each Grade and Socio-*
*metric Position at Grade 12, Contrasted with an Age Curve of Development.*

attenuation) between sociometric position at each grade and at grade 12. It is evident that with the exception of the correlation between grades 10 to 12 the correlations over the four year period are very high and almost constant. That is, there is a high degree of stability in the sociometric positions of the students who were together over the four year period. We are not able to explain why the grade 10 to 12 correlation is lower than the others, especially in view of the fact that it is the only one of the six correlations among the four grades which is lower than .83 (.92 attenuated). It will be noted that the grade 9 to 12 correlation is slightly higher than might be anticipated on the basis of an age curve of development, whereas the grade 11 to 12 correlation is very close to the age curve.

In Table 5.1 we have summarized the results of the various longitudinal studies of sociometric position. It is instructive to note that the majority of studies in which there was little change in the composition of the group and in which male and female students were treated as a single group are as high as might be anticipated on the basis of an age curve of development. The clearest exceptions to this are the correlations for ages 10 to 11, which are significantly lower in both Bonney's (1943) and Staker's (1948) studies.

Where the composition of the group changes considerably over the period of time under consideration, the correlations are somewhat lower than those found in groups with little change in composition. Thus, contrast the +.90 correlation for Bonney's age 9 to 10 group with little change in composition, with the +.68 correlation for another group with greater change in composition. It is likely that much of the stability of sociometric ratings is attributable to a "pecking order" being established early and then maintained as long as the group remains together. One implication of this for schools is the possibility that sociometric position can be altered considerably by rearranging groups of students from one year to another or by the process of in-and-out migration of members of the group.

## Intellectual Achievement

Although educational achievement tests are undoubtedly more accurate measures of the student's intellectual achievement than are the ratings of observers, we have included the correlations on these ratings because they include preschool measures and because they illustrate a pattern of relationship in which sex differences are relatively small.

Kagan and Moss (1962) rated children's behavior before age 10 on

Table 5.1. *Longitudinal Studies of Sociometric Position*

|  |  |  |  | Correlations | | |
|---|---|---|---|---|---|---|
|  |  |  |  | Observed | Theoretically Expected on the Basis of an Age Curve of Development | |
| Author and Date | Sex | N | Ages | A | B<br>When Test Reliability Is Perfect | C<br>Reduced by Actual Test Reliability |
| **Little Change in Composition of Group** |  |  |  |  |  |  |
| Bonney (1943) | M + F | 48 | 8–9 | .84 | .94 | .80 |
|  | M + F | 17 | 9–10 | .90 | .95 | .81 |
|  | M + F | 57 | 10–11 | .67* | .95 | .81 |
| Cannon (1958) | M + F | 17 | 15–18 | .84 | .91 | .77 |
|  | M + F | 39 | 16–18 | .67 | .94 | .80 |
|  | M + F | 69 | 17–18 | .83 | .97 | .82 |
| Staker (1948) | M + F | 50 | 10–11 | .68* | .95 | .81 |
| **Considerable Change in Composition of Group** |  |  |  |  |  |  |
| Bonney (1943) | M + F | 17 | 9–10 | .68 | .95 | .81 |
| Wertheimer† (1957) | M | 100 | 16–18 | .69* | .94 | .80 |
|  | F | 100 | 16–18 | .62** | .94 | .80 |

\* Significant at .05 level.
\*\* Significant at .01 level.
† Although the group formally remained intact in the home room, the students were not likely to be together in other classes and activities.

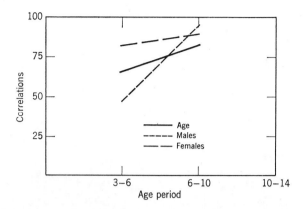

Chart 4. *Correlations (Attenuated) between Ratings of Intellectual Achievement at Each Age Period and Ratings at Ages 10 to 14, Contrasted With an Age Curve of Development.*

*general achievement.* The ratings were based on the child's persistence with challenging tasks, games, and problems and his involvement in activities in which a standard of excellence was applicable. In the period ages 10 to 14, the achievement behavior referred to in Chart 4 was limited to *intellectual achievement.*

In Chart 4 it will be seen that the ratings for girls at each period are highly correlated with the age 10 to 14 ratings and that they are far more consistent than would be expected on the basis of an age curve of development. In the 3 to 6 age period, the correlation of the boys ratings with the criterion is much lower than the girls or the age curve, whereas during the age 6 to 10 period it is almost the same as the girls. We have not included the rating made in the 0 to 3 age period because of the low rater reliability and because in this period the ratings were primarily based on persistence with perceptual-motor activities.

It is evident in Chart 4 that the general achievement pattern is clearly formed before the children entered school and that by ages 3 to 6, two-thirds of the variance for girls in this characteristic at adolescence is already accounted for.

In Table 5.2 we have included additional longitudinal data on ratings of intellectual skills and intellectual competence. The majority of the correlations are equal to or higher than would be estimated from an age curve of development. The clearest exceptions to this are the age 5 to 24 correlation for males and the age 12 to 24 correlation for females. It is noteworthy that, in spite of some shifts in the charac-

Table 5.2. Longitudinal Studies of Intellectual Achievement and Related Characteristics

| | | | | | Correlations | | |
| | | | | | Observed | Theoretically Expected on the Basis of an Age Curve of Development | |
| Author and Date | Variable | Sex | N | Ages | A | B — When Test Reliability Is Perfect | C — Reduced by Actual Test Reliability |
|---|---|---|---|---|---|---|---|
| Peck and Havighurst (1960) | Intellectual skills | M + F | 34 | 10–16 | .78 | .79 | .75 |
| | | M + F | 34 | 13–16 | .92 | .91 | .86 |
| Kagan and Moss (1962) | Intellectual achievement | M | 44 | 5–12 | .39 | .65 | .57 |
| | | M | 44 | 8–12 | .81 | .82 | .72 |
| | | F | 45 | 5–12 | .68 | .65 | .57 |
| | | F | 45 | 8–12 | .76 | .82 | .72 |
| | Achievement versus adult concern with intellectual competence | M | 36 | 5–24 | .13 | .45 | .40 |
| | | M | 36 | 8–24 | .68 | .57 | .53 |
| | | F | 35 | 5–24 | .44 | .45 | .40 |
| | | F | 35 | 8–24 | .49 | .57 | .53 |
| | Intellectual achievement versus adult concern with intellectual competence | M | 36 | 12–24 | .66 | .71 | .65 |
| | | F | 35 | 12–24 | .49 | .71 | .65 |

teristics being rated, the general achievement pattern is a relatively stable characteristic which can be observed by competent raters or measured by means of aptitude and achievement tests (see Chapters 3 and 4). Furthermore, the difference in the stability of this characteristic for males and females is relatively small.

## Aggression

Aggression is a characteristic that has been much emphasized in dynamic theories of personality and in studies of child growth and development. It is an important symptom in personality diagnosis. Furthermore, in many respects it would seem to be one of the more readily observable personality characteristics.

Kagan and Moss (1962) report stability data on several kinds of aggression. Somewhat different results are obtained for each type. Thus physical aggression to peers, although highly stable over the first ten years of life, could not be rated during ages 10 to 14 because of its infrequent occurrence during this period. Aggression toward the mother showed moderate stability for short time intervals only. It is likely that the use of aggression and the forms in which it may be expressed are determined by social norms and expectations. What is permitted at one period in life may be discouraged or prohibited at another period.

One of the more stable forms of aggression is indirect aggression to peers which Kagan and Moss define as "Occurrence of unprovoked nonphysical aggression toward same sex peers, for example, verbal threats or taunts, teasing, destruction or seizure of a peer's property." In Chart 5 we have plotted the correlations for indirect aggression to peers. Insufficient data were available on boys during the first period to warrant a correlation coefficient. However, the correlation for girls between ages 0 to 3 and ages 10 to 14 is much higher than would be expected on the basis of an age curve. This is also true for age periods 3 to 6 versus 10 to 14 for boys. The boys are consistently more stable on this characteristic than are the girls, although the differences are not sizable. This difference is probably due to the greater likelihood of observing the characteristic in boys than in girls and thus providing more evidence for the ratings. In general, by age 6 as much as one-half of the variance in this characteristic at adolescence is already accounted for.

Other longitudinal data on aggression and related characteristics are summarized in Table 5.3. Aggression is clearly established as a characteristic by age 6 and for boys, at least, is a highly stable charac-

Table 5.3. Longitudinal Studies of Aggression and Related Characteristics

| Author and Date | Variable | Sex | N | Ages | Correlations | | |
|---|---|---|---|---|---|---|---|
| | | | | | Observed | Theoretically Expected on the Basis of an Age Curve of Development | |
| | | | | | A | B — When Test Reliability Is Perfect | C — Reduced by Actual Test Reliability [1] |
| Kagan and Moss (1962) | Indirect aggression to Peers | M | 44 | 5–12 | .64 | .65 | .57 |
| | | M | 44 | 8–12 | .71 | .82 | .69 |
| | | F | 45 | 2–12 | .53 | .41 | .35 |
| | | F | 45 | 5–12 | .56 | .65 | .57 |
| | | F | 45 | 8–12 | .60 | .82 | .69 |
| | Agression to mother | M | 44 | 2–12 | −.19** | .41 | .29 |
| | | M | 44 | 5–12 | .23 | .65 | .47 |
| | | M | 44 | 8–12 | .56 | .82 | .71 |
| | | F | 45 | 2–12 | .36 | .41 | .29 |
| | | F | 45 | 5–12 | −.02** | .65 | .47 |
| | | F | 45 | 8–12 | .49* | .82 | .71 |
| | Dominance of peers | M | 44 | 5–12 | .52 | .65 | .55 |
| | | M | 44 | 8–12 | .67 | .82 | .71 |
| | | F | 45 | 5–12 | −.03** | .65 | .55 |
| | | F | 45 | 8–12 | .25** | .82 | .71 |
| | Behavioral disorganization | M | 44 | 2–12 | .35 | .41 | .28 |
| | | M | 44 | 5–12 | .52 | .65 | .51 |
| | | M | 44 | 8–12 | .67 | .82 | .66 |
| | | F | 45 | 2–12 | −.02* | .41 | .28 |
| | | F | 45 | 5–12 | −.03** | .65 | .51 |
| | | F | 45 | 8–12 | .25** | .82 | .66 |

| | | | | | | |
|---|---|---|---|---|---|---|
| Indirect aggression to peers versus adult competitiveness | M | 36 | 2-24 | .31 | .28 | .24 |
| | M | 36 | 5-24 | .56 | .45 | .40 |
| | M | 36 | 8-24 | .51 | .57 | .52 |
| | M | 36 | 12-24 | .45 | .71 | .65 |
| | F | 35 | 2-24 | .22 | .28 | .24 |
| | F | 35 | 5-24 | .18 | .45 | .40 |
| | F | 35 | 8-24 | .07** | .57 | .52 |
| | F | 35 | 12-24 | —.03** | .71 | .65 |
| Competitiveness versus adult competitiveness | M | 36 | 5-24 | .51 | .45 | .30 |
| | M | 36 | 8-24 | .39 | .57 | .50 |
| | F | 35 | 5-24 | .52 | .45 | .30 |
| | F | 35 | 8-24 | .08** | .57 | .50 |
| Tuddenham (1959) Aggression | M | 32 | 16-36 | .68 | .66 | .49 |
| | F | 38 | 16-36 | .07** | .66 | .50 |
| Jersild (1935) Number of conflicts | M & F | 15 | 2.5-3.5 | .63 | .84 | .68 |
| | M & F | 13 | 3.5-4.5 | .69 | .88 | .71 |

* Significant at .05 level.
** Significant at .01 level.

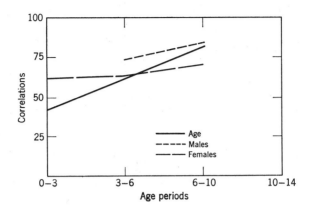

*Chart 5.  Correlations (Attenuated) between Ratings of Aggression at Each Age Period and at Ages 10 to 14, Contrasted with an Age Curve of Development.*

teristic.  Thus of 17 observed correlations for males in which the same characteristic was rated at each time period, 12 were as high or higher than the age curve values.  In contrast, for the females 13 out of 18 correlations were *below* the age curve values.  In 17 comparisons of males versus females, 15 of the correlations are *higher* for the males.

It is not clear whether the higher correlations for males in comparison with females are attributable to the greater stability of this characteristic for males or to the greater ease of detecting and rating this characteristic in males than in females.  Aggressiveness is usually frowned upon in girls and especially in women, whereas it may be regarded as natural in boys and, in some ways, viewed as a characteristic related to manliness and the ability to take care of oneself.  Whether the socialization process actually does change the basic characteristic or merely alter the forms in which it may be expressed is a problem for future research.  We are of the opinion that some of the difference between the stability of aggression in males and females is attributable to the difficulty that observers and raters have in detecting the more subtle forms of aggression in the behavior of females as compared with the more easily observed forms of this behavior in males.

This view is, in part, supported by the studies involving adults where the stability is relatively high for males but almost completely disappears for females.  In spite of some change in the characteristics rated in the adults as compared with the characteristics rated in the earlier years, most of the observed correlations for males are as high, or higher, than the values estimated from the age curves, while the reverse is true for females.

## Dependence-Passivity

A set of characteristics which can be observed and rated with a substantial level of rater agreement includes dependence on others for support and direction and passive response to frustration and difficulties.   These are of special interest to us here because of the way in which the culture may define the desired role for males in contrast to females with regard to dependence and passivity.

Kagan and Moss (1962) have included several measures of passivity and dependence in their longitudinal study.   In addition, several other studies have obtained longitudinal data on closely related characteristics.   One of the more stable characteristics in this set is *passive reaction to frustration*.   This is a rating of "the degree to which the child acquiesced or withdrew in the face of attack or frustrating situations, in contrast to an active attempt to overcome and deal with environmental frustrations."   During the early years, passivity was manifested through behaviors such as (1) retreat when dominated by sibling, (2) no reaction when goal object is lost, (3) withdrawal when blocked from goal by environmental obstacle, and (4) withdrawal from mildly noxious or potentially dangerous situations.   During the school years, the passivity rating was based primarily on (1) withdrawal to attack or social rejection, and (2) withdrawal from difficult and frustrating task situations.

In Chart 6 it may be seen that *passive reaction to frustration* is a highly stable characteristic for girls which is highly consistent from

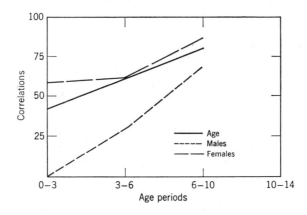

*Chart 6.   Correlations (Attenuated) between Ratings of Dependency at Each Age Period and at Ages 10 to 14, Contrasted with an Age Curve of Development.*

Table 5.4. Longitudinal Studies of Dependency and Related Characteristics

| Author and Date | Variable | Sex | N | Ages | Correlations | | |
|---|---|---|---|---|---|---|---|
| | | | | | Observed | Theoretically Expected on the Basis of an Age Curve of Development | |
| | | | | | A | B When Test Reliability Is Perfect | C Reduced by Actual Test Reliability |
| Kagan and Moss (1962) | Conformity | M | 44 | 2–12 | .27 | .41 | .32 |
| | | M | 44 | 5–12 | .16** | .65 | .57 |
| | | M | 44 | 8–12 | .53*** | .82 | .75 |
| | | F | 45 | 2–12 | −.24** | .41 | .32 |
| | | F | 45 | 5–12 | .35* | .65 | .57 |
| | | F | 45 | 8–12 | .68 | .82 | .75 |
| | Passivity | M | 44 | 2–12 | .00** | .41 | .30 |
| | | M | 44 | 5–12 | .24* | .65 | .53 |
| | | M | 44 | 8–12 | .60** | .82 | .77 |
| | | F | 45 | 2–12 | .44 | .41 | .30 |
| | | F | 45 | 5–12 | .51 | .65 | .53 |
| | | F | 45 | 8–12 | .76 | .82 | .77 |
| | Passivity versus adult withdrawal from stress | M | 36 | 2–24 | −.06 | .28 | .17 |
| | | M | 36 | 5–24 | .06 | .46 | .32 |

| | N | Age | | | |
|---|---|---|---|---|---|
| M | 36 | 8–24 | .27 | .57 | .43 |
| M | 36 | 12–24 | .36 | .71 | .53 |
| F | 35 | 2–24 | .22 | .28 | .17 |
| F | 35 | 5–24 | .26 | .46 | .32 |
| F | 35 | 8–24 | .48 | .57 | .43 |
| F | 35 | 12–24 | .67 | .71 | .53 |
| M | 36 | 2–24 | .05 | .28 | .18 |
| M | 36 | 5–24 | .31** | .46 | .64 |
| M | 36 | 8–24 | .25 | .57 | .45 |
| F | 35 | 2–24 | .29 | .28 | .18 |
| F | 35 | 5–24 | .35* | .46 | .64 |
| F | 35 | 8–24 | .57 | .57 | .45 |
| M | 44 | 5–12 | −.33** | .65 | .59 |
| M | 44 | 8–12 | .52 | .82 | .63 |
| F | 45 | 5–12 | .64 | .65 | .59 |
| F | 45 | 8–12 | .65 | .82 | .63 |
| M | 36 | 2–24 | .00 | .28 | .18 |
| M | 36 | 5–24 | −.03** | .46 | .64 |
| M | 36 | 8–24 | .05** | .57 | .46 |
| M | 36 | 12–24 | .22** | .71 | .56 |
| F | 35 | 2–24 | .25 | .28 | .18 |
| F | 35 | 5–24 | .24** | .46 | .64 |
| F | 35 | 8–24 | .55 | .57 | .46 |
| F | 35 | 12–24 | .54 | .71 | .56 |

Independence† versus adult withdrawal from stress

Passivity versus adult dependency in vocational choice

Table 5.4. Longitudinal Studies of Dependency and Related Characteristics (Continued)

| Author and Date | Variable | Sex | N | Ages | Observed A | Correlations — Theoretically Expected on the Basis of an Age Curve of Development | |
|---|---|---|---|---|---|---|---|
| | | | | | | B When Test Reliability Is Perfect | C Reduced by Actual Test Reliability |
| Peck and Havighurst (1960) | Emotional independence | M + F | 34 | 10–16 | .79 | .79 | .74 |
| | | M + F | 34 | 13–16 | .95** | .90 | .85 |
| Macfarlane et al. (1954) | Emotional independence | M + F | 41 | 7–14 | .34 | .71 | .49 |
| | | M + F | 41 | 9–14 | .51 | .80 | .58 |
| | | M + F | 41 | 11–14 | .66 | .88 | .63 |
| | Shyness | M + F | 41 | 7–14 | .60 | .71 | .43 |
| | | M + F | 41 | 9–14 | .54 | .80 | .50 |
| | | M + F | 41 | 11–14 | .73* | .88 | .55 |
| Tuddenham (1959) | Self-sufficiency | M | 19 | 16–36 | .08 | .66 | .32 |
| | | F | 17 | 16–36 | .55* | .66 | .36 |

* Significant at .05 level.
** Significant at .01 level.
† Signs of correlations reversed because of scales used.

ages 0 to 3 on insofar as the correlation with this form of passivity at ages 10 to 14 is concerned.    In marked contrast is the lack of stability of this characteristic in boys in the early childhood periods. Only at ages 6 to 10 versus 10 to 14 does the relationship for boys approach the level of relationship that might be expected on the basis of an age curve.

In Table 5.4 we have summarized the findings on this and related characteristics from a number of studies.    The results are very similar for most of these characteristics.    In 20 comparisons, the stability values for females are higher than for males in all except one.

In the comparisons of the observed correlations with those that might be expected on the basis of an age curve, the observed values for males are *lower* than the expected values in 20 out of 20 instances. In contrast, in only 6 out of 20 instances are the observed values for the females lower than the age curve values.

Where both sexes are combined, as in the Macfarlane study, the values are not very different from the average of the values of the two sexes taken separately.    The Peck and Havighurst values for both sexes combined in the period ages 13 to 16 are especially high— higher than would be expected for females alone.

The difference between the sexes in the stability of dependence and passivity is very clear and is in marked contrast with the difference between the sexes in aggression.    Here again, we can only speculate about the causes of these differences.    It would seem highly likely that just as aggressiveness is permitted, and even encouraged, in males, so passivity and dependence are regarded as peculiarly female characteristics.    It is not likely that females in the United States would be encouraged to be independent, self-sufficient, or overly active and retaliatory in dealing with frustrating situations.    There seems to be little doubt that, in general, females appear to exhibit more passivity and dependence than do males, at least during adolescent and later stages of development.

However, we do not know for certain whether males and females are really different in this respect or whether the longitudinal findings are more nearly symptoms of the different forms that this characteristic takes in males and females and the difficulties of observing and rating this characteristic in males as contrasted with females.    Kagan and Moss report similar correlations for girls and boys during the age period 0 to 10, whereas the correlations are very different for the two sexes after age 10.    We are inclined to the view that more subtle and indirect techniques of measuring passivity and dependence will eventually reveal very similar results for males and females although the overt manifestations of these characteristics will continue to show marked sex differences.

## Summary (Observational Material)

In Chart 7 we have contrasted the longitudinal consistency of selected characteristics with the values based on an age curve.   It is clear from this chart that each of the characteristics shows more consistency in the early years than would be expected on the basis of an age curve.   The correlation values for these selected characteristics for at least one of the sexes are such that over one-third of the variance during adolescence is predictable as early as ages 0–3.   This is in harmony with theoretical formulations about the importance of the early years for personality development.

The difference between boys and girls for these characteristics may, as we have pointed out earlier, be attributable to the way in which male and female roles are emphasized in this society such that aggression is discouraged in females, whereas dependence is discouraged in males.   There is little doubt that the majority of males and females learn their respective sex roles by adolescence.   Some support for this may be found in the intercorrelations reported by Kagan and Moss (1962, pp. 53, 88) among the pre-adolescent measurements which are very similar for males and females in contrast to the correlations with the adolescent values that are so different for the two sexes.

However, quite another inference may be made about these relationships.   The observations, which are translated into ratings, are made on the overt behaviors of the subjects.   As such, they are indications

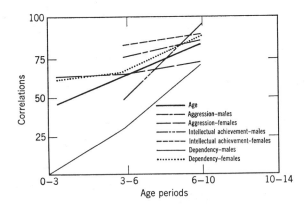

Chart 7.   *Correlations (Attenuated) between Ratings of Selected Characteristics at Each Age Period and at Ages 10 to 14, Contrasted with an Age Curve of Development.*

of the outward manifestations of personality rather than the more deep-seated characteristics.  It is quite possible that individuals do learn the appropriate behaviors for each age period and for the role required by our society—and that it is the variations in these learned behaviors which are represented by the observations and ratings reported in the foregoing.  The more deep-seated characteristics would be displayed only in the subtlest ways, if they are discouraged by the society.  Thus the difference between the males and females in the stability of aggression and dependence may be artifacts of the methods of observations and ratings used in these studies.  It is likely that only through more sensitive methods of measuring personality characteristics will the full stability of these deeper aspects of personality be revealed.


CHARACTERISTICS MEASURED BY METHODS WHICH MINIMIZE
CONSCIOUS CONTROL BY SUBJECTS

In the previous section we have pointed to the way in which observational data may reveal the stability (or lack of it) of the more overt manifestations of personality characteristics.  It was suggested in that section that more deep-seated aspects of personality may not be detected by observational techniques with very much precision and that some of the sex and age differences may disappear if the measurement techniques probe some of the less conscious aspects of personality.

Some techniques have been devised to measure personality characteristics by means of tasks and situations in which the subject may not be fully aware of what he is revealing when he responds.  Such techniques include perceptual-cognitive tasks in which the subject reveals much about himself by the way in which he approaches the tasks and the "solutions" he gives to the tasks as he has perceived them.  Also included here are the projective and semiprojective techniques such as the Rorschach test, the Thematic Apperception test, sentence-completion tests, and various drawing tests interpreted as projective instruments.  In these techniques, the subject responds as best he can without comprehending the possible ways in which his responses may be interpreted.

Although a very sophisticated subject may, in part, be aware of some of the ways in which his responses may be interpreted, it is likely that the majority of subjects will not fully comprehend what meaning can be attached to their responses.  It is also likely that the majority of the subjects will not be able to consciously control their responses in such a way as to give what they believe to be a favorable picture of themselves to the examiner.

Perhaps the major source of error and difficulty with these instruments is in determining exactly what is being measured by these indirect techniques. Although the responses of the subject may be clear and recordable, the inferences to be drawn from the responses are less clear. The examiners may differ greatly in the *interpretations* they give to a set of responses.

In the following, we will attempt to summarize some of the longitundinal studies which have been made with perceptual-cognitive tests as well as the situation that prevails with longitudinal studies using projective and semiprojective instruments.

## Perceptual-Cognitive Tasks

Witkin (1962) and his associates have for some time been using experimental techniques to measure the judgments subjects make about the position of objects when the field or context in which such objects are normally seen is removed. Although the experiments differ, an illustration of one of them would be the angle at which a lighted rod is seen when the rest of the room is so dark that the subject cannot use these context clues. Witkin relates field dependence to more basic personality characteristics.

In Table 5.5 we report some of the longitudinal studies on field dependence from ages 10 to 17, and from ages 17 to 21. It will be noted that with only one exception the observed values are not significantly different from the values based on an age curve of development. Although we are not entirely clear what is being measured by this technique, we believe that some aspects of ego development are probably involved in these rather complex but precise experiments. It is of interest to note that the earlier measurements, ages 10 to 17, are as close to the estimated values as are the later measurements, ages 18 to 21. We are able to offer no explanation for some of the sex differences in the correlations.

We have also included in Table 5.5 Witkin's study of the results of a drawing test given to the same subjects at ages 10 to 17. These were scored for the sophistication of body concepts revealed in the drawings. Although it is clear that the sophistication of body concept does change with age, the rank order of the individuals and their drawings at the two different ages are quite consistent and are very close to the correlation anticipated on the basis of an age curve of development.

Table 5.5. *Longitudinal Studies of Perceptual-Cognitive Development*
*(Adapted from Witkin, et al. 1962)*

| | | | | | | Correlations | |
| | | | | | | Theoretically Expected on the Basis of an Age Curve of Development | |
| | | | | | A | B | C |
| Author and Date | Variable | Sex | N | Ages | Observed | When Test Reliabity Is Perfect | Reduced by Actual Test Reliability |
|---|---|---|---|---|---|---|---|
| Witkin | Field dependence | M | 27 | 10–14 | .64 | .84 | .76 |
| | | M | 27 | 10–17 | .50 | .77 | .69 |
| | | M | 27 | 14–17 | .87 | .91 | .82 |
| | | F | 24 | 10–14 | .88 | .84 | .76 |
| | | F | 24 | 10–17 | .79 | .77 | .69 |
| | | F | 24 | 14–17 | .94** | .91 | .82 |
| Flugel | | M | 13 | 17–20 | .85 | .92 | .83 |
| | | F | 14 | 17–20 | .80 | .92 | .83 |
| Barman | | M | 32 | 18–21 | .91 | .93 | .84 |
| | | F | 30 | 18–21 | .74 | .93 | .84 |
| Witkin | Sophistication of body concept | M | 14 | 10–17 | .73 | .77 | .69 |

** Significant at .01 level.

159

## Projective Techniques

We had hoped to secure a large number of longitudinal studies utilizing projective techniques. Although we have been able to find a number of follow-up studies using Rorschach, TAT, and sentence completion tests, we can find very few studies in which the data have been quantitatively treated from a longitudinal point of view. All too frequently, the data have been used for case studies and normative purposes rather than for the determination of stability and change. We hope that further analysis of these data in longitudinal terms will make it possible to determine the extent to which some of the rather complex efforts to get at the nonconscious thoughts of persons and the deeper aspects of personality reflect the stability that we find in the more surface manifestations of personality and character.

### CHARACTERISTICS MEASURED BY SELF-REPORT TECHNIQUES

Self-report techniques have been used very widely in measuring interests, attitudes and values, and personality characteristics. For the most part, these techniques include a questionnaire, check list, or collection of questions that the examinee attempts to answer. Although some of the questions may get at deep-seated aspects of personality in such a way that the subject has little awareness of what is being revealed, for the most part the questions are likely to be ones for which the subject can determine the favorable response. Thus a major source of error may be the variations in the "truthfulness" of the examinee's responses. However, it is also possible that many of the questions attempt to get at aspects of behavior in such a way that the subject has serious difficulty in providing an accurate self-description. Thus the "accuracy" of the responses are subject to conscious as well as nonconscious sources of error.

From the longitudinal research point of view, two sources of error may mask the stability that may really be present. The first source of error may be the different motivation the subject may have in responding honestly at the two different points in time. If he answers it honestly and carefully at one time and then attempts to produce a particular picture of himself or answers it carelessly at another time, the results will show lower correlations than might be expected. This would be especially true if the honesty and carefulness in responding vary for the same subject at the different test administrations, or if the group of subjects varies considerably in their honesty and carefulness at either administration of the test.  A second source of error arises from the appropriateness of the questions for subjects at the two ages under consideration.  If the questions are appropriate for adults and the first administration is in the adolescent period, the level of stability will be lower than expected because the meaning of the questions for the examinee changes from one administration to the next.

In general, we would expect less stability in self-report measurements than in either the observational techniques or the methods which minimize conscious control by the subjects.  There are so many uncontrolled sources of errors in the self-report techniques which cannot be accounted for simply by corrections for unreliability.  It is also unfortunate that the nature of the self-report techniques for which longitudinal data are available are primarily those which have been developed for secondary school and college students.  Thus our longitudinal data must be limited to the adolescent and later years.

## Interests—Kuder Preference Record

The Kuder Preference Record and the Strong Vocational Interest Blank are the two interest questionnaries that have most frequently been used in longitudinal research.  Both these instruments were primarily intended for the high school and college student and this is reflected in the language used in the items, the content of the items, and the normative data.  Both instruments are intended for guidance purposes in relation to choice of a field of professional specialization.

The Kuder Preference Record consists of a large number of activities which are presented in sets of three which the individual is to mark to indicate the activity he likes most and the activity he likes least.  The entire instrument attempts to measure the relative preferences the individual has for clusters of items classified as mechanical, computational, scientific, etc.

We have selected only two interest areas for attention here.  These two areas, *mechanical* and *artistic*, have relatively high reliabilities and are of special interest to us because they are likely to show some differences between the sexes and because the activities represented in the two areas are likely to be widespread throughout the society and to be engaged in by individuals at all age levels from adolescence on.

In Chart 8 we have shown the correlations (attenuated) between the scores at each age and age 17.  It is evident that, with one exception, the level of stability is lower than would be expected on the basis of an age curve of development.  This may, in part, be explained on the

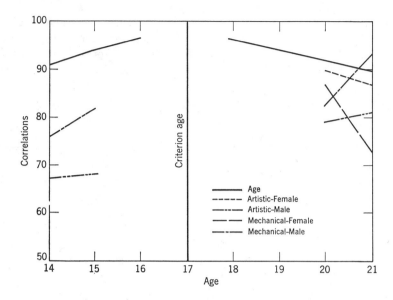

*Chart 8. Correlations (Attenuated) between Scores on the Kuder Preference Record at Selected Ages and at Age 17, Contrasted with an Age Curve of Development.*

basis of the homogeneity of the samples studied which consist primarily of high school graduates or college students. Of some interest here is the slope of the curves before age 17 as contrasted with those after age 17. Before age 17 there appears to be very rapid development and change, while after this age the interests appear to stabilize such that there is as much consistency over a four year period (after 17) as there is over a one or two year period (before 17). This would suggest that major changes in the interests sampled by the Kuder Preference Record are likely to take place in the high school and early college years.

In Table 5.6 we have summarized the longitudinal findings on these two interest scores and have contrasted them with estimates based on an age curve of development. It will be seen that wherever the males and females are treated separately, stability values observed tend to be slightly higher for males on the mechanical score, whereas they tend to be slightly higher for the females on the artistic score. These results are in harmony with sex role expectations in our society with respect to mechanical and artistic activities.

Towner (1948) identified students who were veterans of the armed services and contrasted them with students who had not been in military service.  Although these two groups of Towner's do differ from each other in the stability values, this seems to be little more than a reflection of the age periods involved.  When each group is compared with other groups over the same age period, there seems to be little difference between veterans and nonveterans.  Bordin and Wilson (1953) compared male college students who changed their college curricula with those who did not.  On both scores, the students who did not change curricula had higher stability correlations than did the students who changed curricula.  This would suggest that some individuals have much more stable interests than do others (whether the stability of interests is attributable to the individual's characteristics or the change of curriculum is a matter of conjecture) and that future research might be devoted to an understanding of the individual as well as environmental characteristics which make for stability and change in interest patterns.

The low correlations between scores from ages 14 to 18 in contrast with the markedly higher correlations between ages 17 and above may be interpreted in two ways.  It is possible that marked changes take place in interests between ages 14 and 17 and that this is the point in the adolescent period where individuals make the greatest change in determining the activities they prefer and the social and occupational roles they desire.  However, another possibility is that the language and content of the Kuder items are less appropriate to the 16 year and earlier levels than it is to the later ages.  It is to be hoped that future research will determine whether broad interest patterns are developed before age 16, whereas more specific interests in relation to particular activities and occupations are developed throughout early adulthood.  Undoubtedly, some of the items in the Kuder instrument are highly age-related in language and/or content, whereas other items are less changeable over long periods of time.

Herzberg's et al. (1954) study shows about the same levels of stability for these two interest areas for students in college in the age period 17 to 20 as for persons who were out of school during this age period.  This does raise questions about the effect of college experiences on these two interests areas.

About one-half the observed correlations are significantly below those which might have been expected if interests developed in accordance with a simple age curve of development.  The median difference is about .11 correlation points when the comparison is made between the observed correlations and the age curve values reduced to account for unreliability.

The simplest explanation for the lower observed correlations is to

Table 5.6. Longitudinal Studies of Interests—Kuder Preference Record

| | | | | Observed | Theoretical Correlations Based on an Age Curve of Development | |
|---|---|---|---|---|---|---|
| | | | | | When Test Reliability Is Perfect | Reduced by Actual Test Reliability |
| Author and Date | Sex | N | Ages | A | B | C |
| Artistic | | | | | | |
| Long and Perry (1953) | M | 32 | 19–22 | .66* | .93 | .85 |
| Wright and Scarborough (1958) | M | 174 | 19–20 | .67** | .97 | .88 |
| | M | 125 | 19–22 | .72*** | .93 | .85 |
| | F | 205 | 19–20 | .85 | .97 | .88 |
| | F | 105 | 19–22 | .74** | .93 | .85 |
| Bordin and Wilson (1953) | M | 165 | 19–20[a] | .83 | .97 | .88 |
| | M | 91 | 19–20[b] | .70** | .97 | .88 |
| Herzberg et al. (1954) | M | 101 | 17–20[e] | .72*** | .92 | .84 |
| | F | 48 | 17–20[e] | .82 | .92 | .84 |
| | M | 49 | 17–20[f] | .80 | .92 | .84 |
| | F | 74 | 17–20[f] | .77 | .92 | .84 |
| Reid (1951) | M + F | 145 | 18–19½ | .76** | .96 | .87 |
| Rosenberg (1953) | M | 91 | 14–17 | .61** | .91 | .83 |
| | F | 86 | 14–17 | .69* | .91 | .83 |
| Herzberg and Bouton (1954) | M | 62 | 17–21 | .74 | .90 | .82 |
| | F | 68 | 17–21 | .79 | .90 | .82 |
| Stoops (1953) | M | 45 | 14–16 | .74* | .94 | .86 |
| | F | 64 | 14–16 | .65** | .94 | .86 |

Correlations

164

| Study | Sex | N | Age | | | |
|---|---|---|---|---|---|---|
| Towner (1948) | M | 85 | 16-18[c] | .62** | .94 | .86 |
| | M | 26 | 16-20[d] | .55* | .89 | .81 |
| Adams (1957) | M + F | 57 | 14-17 | .63** | .91 | .83 |
| Jacobs (1949) | M + F | 76 | 14-16 | .82 | .94 | .86 |
| Mechanical | | | | | | |
| Long and Perry (1953) | M | 32 | 19-22 | .14** | .93 | .84 |
| Wright and Scarborough (1958) | M | 174 | 19-20 | .78*** | .97 | .87 |
| | M | 125 | 19-22 | .69*** | .93 | .84 |
| | F | 205 | 19-20 | .72*** | .92 | .87 |
| | F | 105 | 19-22 | .71** | .93 | .84 |
| Bordin and Wilson (1953) | M | 165 | 19-20[a] | .86 | .97 | .84 |
| | M | 91 | 19-20[b] | .78 | .93 | .84 |
| Herzberg et al. (1954) | M | 101 | 17-20[e] | .75* | .92 | .83 |
| | F | 48 | 17-20[e] | .78 | .92 | .83 |
| | M | 49 | 17-20[f] | .74 | .92 | .83 |
| | F | 74 | 17-20[f] | .74 | .92 | .83 |
| Reid (1951) | M + F | 145 | 18-19½ | .84 | .96 | .86 |
| Rosenberg (1953) | M | 91 | 14-17 | .68** | .91 | .82 |
| | F | 86 | 14-17 | .53** | .91 | .82 |
| Herzberg and Bouton (1954) | M | 62 | 17-21 | .84 | .90 | .81 |
| | F | 68 | 17-21 | .66* | .90 | .81 |
| Stoops (1953) | M | 45 | 14-16 | .79 | .94 | .85 |
| | F | 64 | 14-16 | .55** | .94 | .85 |
| Towner (1948) | M | 85 | 16-18[c] | .74* | .94 | .85 |
| | M | 26 | 16-20[d] | .82 | .89 | .80 |
| Adams (1957) | M + F | 57 | 14-17 | .72 | .91 | .82 |
| Jacobs (1949) | M + F | 76 | 14-16 | .79 | .94 | .85 |

* Significant at .05 level.
** Significant at .01 level.

[a] Unchanged curriculum
[b] Changed curriculum
[c] Nonveterans
[d] Veterans
[e] Work follow-up
[f] College follow-up

be found in the variability of the samples.   Since these are high school graduates and college students, they are not as variable on these interests as is the normative population reported in the Kuder technical manual.   It is quite likely that if an allowance is made for the reduced variability of the samples, the median differences between these estimates and the observed correlations would be somewhat smaller.

However, we do not find a level of stability that is characteristically higher than might be expected on the basis of an age curve.   It is possible that the age period covered by these longitudinal data, 14 to 22, is the period of greatest change in interests in vocational activities. This is the time when a large portion of activities in the Kuder instrument are first encountered and tried, for example,

> Interview people in surveys of public opinion.
> Interview an employer for a job.
> Write a letter applying for a job.
> Study literary criticism.
> Study the history of philosophy.
> Be a social service visitor.
> Read about how language is changing.
> Win a fellowship to study educational methods in this country.
> Read an article on how to treat mental ills.

At least for these activities it is reasonable to expect considerable change as individuals try, select, and discard specific activities.

It is possible that a broader approach to interests with fewer age-related activities and with a language and form appropriate to younger ages may reveal earlier stabilization of general interests than is revealed by these data based on the Kuder Preference Record.

Finally, it is possible that stability of interests will be greater if the same type of broad interests is measured by a series of instruments and questions appropriate to each age group.   Just as the Stanford-Binet gets at intelligence by using different problems and questions at each level, so it may be possible to develop a sequential set of interest instruments which may reveal greater consistency from one level to another than is evidenced by a single instrument and set of items for all ages from 14 and up.

### Interests—Strong Vocational Interest Blanks

The stability of the Strong Vocational Interest Blank has been investigated over longer periods of time than have any of the other instruments except for the intelligence tests.   This Blank consists of 400 items to which the examinee indicates like, indifference, and dislike.   The items are scored on the basis of a weighting scheme for each occupation based on the responses of a group of men successfully engaged in that occupation.

This instrument was constructed primarily for college students, and most of the occupational keys are for professions where a college degree is a minimal requisite.

Table 5.7 summarizes the longitudinal findings for a few of the occupational keys for the Strong Blank.   It is evident that in the period before age 18 or entrance to college, the stability of this instrument is much below that to be expected on the basis of an age curve of development.   Here again we may infer that this is the period of rapid development of vocational interests or that the instrument is less appropriate in language and content for these earlier ages than it is for the post high school or college years.

After age 20, the observed correlations tend to be as high or higher than the age estimate values and there is little reduction in the observed correlations for varying periods of time after age 22.   The correlations for a five-year period (ages 22 to 27) are not much greater than the correlations over 10 or 15 year intervals.*   Thus it is highly likely that vocational interests as measured by this instrument are highly stabilized by the end of the college years and remain relatively constant thereafter.

Strong has also compared the pattern of interest profiles over periods as long as 22 years.   Here again, the overall pattern remains relatively constant over long periods of time.   However, these are median values, and Strong points out that for some individuals there are remarkably high levels of stability, whereas for other individuals there is almost no correlation between the interest patterns on the two measurements. Further research is needed to identify the individual and/or environmental characteristics which make for high as well as low levels of stability.   Research is also needed to determine which items in the Strong VIB are age related, which items are highly changeable over time, and which are highly stable.

The point of all this is that we do not have a clear picture of the stability of interests.   Roe (1957) has proposed a model of general interests and has theorized about the early development of such inter-

---

* The Kelly results do show as much stability as would be expected from an age curve of development, but are considerably lower than the results from other studies over comparable age periods.   This may be accounted for by the combination of males and females and the conditions of test administration.

Table 5.7. Longitudinal Studies of Interests—Strong Vocational Interest Blank

| | | | | | Correlations | | |
|---|---|---|---|---|---|---|---|
| | | | | | | Theoretically Expected on the Basis of an Age Curve of Development | |
| | | | | Observed | | When Test Reliability Is Perfect | Reduced by Actual Test Reliability |
| Author and Date | Sex | Ages | N | A | | B | C |
| Lawyer | | | | | | | |
| Stordahl (1954) | M | 18–20 | 111 | .73** | | .95 | .86 |
| | M | 18–20 | 70 | .73*** | | .95 | .86 |
| Van Dusen (1940) | M | 18–21½ | 76 | .50** | | .92 | .84 |
| Strong (1934) | M | 22–27 | 223 | .77 | | .90 | .82 |
| Channing et al. (1941) | M | 16–18 | 64 | .59** | | .94 | .86 |
| Taylor (1942) | M | 16–20 | 64 | .57** | | .89 | .81 |
| Burnham (1942) | M | 18–21 | 188 | .69** | | .93 | .85 |
| Powers (1956) | M | 34–44 | 109 | .77 | | .88 | .80 |
| Kelly (1955) | M + F | 25–45† | 368 | .65 | | .74 | .67 |
| Engineer | | | | | | | |
| Stordahl (1954) | M | 18–20 | 111 | .78** | | .95 | .90 |
| | M | 18–20 | 70 | .79** | | .95 | .90 |
| Van Dusen (1940) | M | 18–21½ | 76 | .85 | | .92 | .87 |
| Strong (1934) | M | 22–27 | 223 | .84 | | .90 | .86 |
| Burnham (1942) | M | 18–21 | 188 | .78** | | .93 | .88 |
| Powers (1956) | M | 34–44 | 109 | .82 | | .88 | .84 |

| | | N | | | |
|---|---|---|---|---|---|
| Strong (1952) | M | 19–20 | 247 | .91 | .97 | .92 |
| | M | 19–29 | 185 | .77 | .81 | .77 |
| | M | 19–38 | 203 | .76** | .71 | .67 |
| Kelly (1955) | M + F | 25–45‡ | 368 | .65 | .74 | .67 |
| Chemist | | | | | | |
| Channing et al. (1941) | M | 16–18 | 64 | .65** | .94 | .84 |
| Taylor (1942) | M | 16–20 | 64 | .56** | .89 | .79 |
| Interest Profile | | | | | | |
| Strong (1951) | M | 17–19 | 57 | .81* | .94 | .89 |
| | M | 20–28 | 50 | .72 | .84 | .80 |
| | M | 22–27 | 50 | .84 | .90 | .86 |
| | M | 22–44 | 228 | .75* | .71 | .67 |

\* Significant at .05 level.
\*\* Significant at .01 level.
† Minister score.
‡ Mathematician score.

ests.    In contrast, interests in particular occupations and particular activities may be developed relatively late, perhaps during the high school and college years.    Further research is needed to determine the interests which are developed very early in life and those which are developed much later.

## Values and Attitudes

The Allport-Vernon Study of Values has been given to large numbers of college students.    Longitudinal data have been gathered on this test for periods of from one to 20 years.    Like the Strong and Kuder tests, this is a self-report questionnaire in which the examinee may give as accurate a report as he can of his views and attitudes and he may consciously or (unconsciously) bias his report in many ways.

One feature of the Allport-Vernon (also of the Kuder) is the forced choice character of the responses such that the individual who is high on one category must be lower on another category since he is indicating his preference for one value over another.

In Table 5.8, we have listed the correlations for two of the Allport-Vernon scales—religion and aesthetics.    These are the two with the highest reliability figures.    It will be seen that these observed values approach the values based on an age curve although they are characteristically about .11 below the estimated values.    Although the observed correlations for the Study of Values do tend to follow the results estimated on the basis of an age curve, the majority of the correlations are significantly lower than the estimated values.    It is difficult to determine whether this is due to actual changes in values, especially during the college years, or whether the lower correlations are a function of the special characteristics of the samples used and of the instrument.

In Table 5.8 we have also included the results of a number of longitudinal studies of attitudes and opinions.    In almost every instance these observed values are very much below the estimated values based on an age curve of development as well as those reported for the Allport-Vernon instrument.    It is likely that the instruments represented in Table 5.8 include attempts to get at basic values, general attitudes, as well as more specific opinions.    We believe Darley (1938, p. 198) best summarized the situation in the following paragraph.

If "attitudes" and "adjustments" be considered, for example, as "opinions," and if these "opinions" be further considered as directed toward self and/or toward objects (or systems) external to self, they may be visualized as falling on a continuum of stability.    "Opinions" recently arrived at in regard to

Table 5.8. Longitudinal Studies of Values, Attitudes, and Related Characteristics

| Author and Date | Variable | Sex | N | Ages | Correlations | | |
| | | | | | Observed | Theoretically Expected on the Basis of an Age Curve of Development | |
| | | | | | A | B When Test Reliability Is Perfect | C Reduced by Actual Test Reliability |
|---|---|---|---|---|---|---|---|
| Whitely (1938) | Allport-Vernon-Religious | M | 84 | 18–21 | .66* | .93 | .78 |
| | | M | 84 | 19–21 | .68* | .94 | .79 |
| | | M | 84 | 20–21 | .74 | .97 | .81 |
| Todd (1941) | | M + F | 94 | 17–20 | .64* | .92 | .77 |
| Kelly (1955) | | M + F | 368 | 25–45 | .60 | .75 | .63 |
| Burgemeister (1940) | | F | 164 | 18–19 | .74* | .97 | .82 |
| Tyler (1961) | | M | 354 | 18–21 | .67** | .93 | .78 |
| | | F | 164 | 18–21 | .71* | .93 | .78 |
| Whitely (1938) | Allport-Vernon-Aesthetic | M | 84 | 18–21 | .45* | .93 | .78 |
| | | M | 84 | 19–21 | .41** | .94 | .79 |
| | | M | 84 | 20–21 | .55** | .97 | .81 |
| Todd (1941) | | M + F | 94 | 17–20 | .61** | .92 | .77 |
| Kelly (1955) | | M + F | 368 | 25–45 | .53* | .75 | .63 |
| Burgemeister (1940) | | F | 164 | 18–19 | .74* | .97 | .82 |
| Tyler (1961) | | M | 354 | 18–21 | .60** | .93 | .78 |
| | | F | 164 | 18–21 | .72 | .93 | .78 |

Table 5.8. Longitudinal Studies of Values, Attitudes, and Related Characteristics (Continued)

| | | | | | Correlations | | |
| | | | | | | Theoretically Expected on the Basis of an Age Curve of Development | |
| | | | | | A | B | C |
| Author and Date | Variable | Sex | N | Ages | Observed | When Test Reliability Is Perfect | Reduced by Actual Test Reliability |
|---|---|---|---|---|---|---|---|
| Corey (1940) | Attitude toward church | F | 100 | 18–19 | .55** | .97 | .80 |
| Whistler (1940) | Attitudes towards religious revival | M + F | 27 | 18–19 | .30** | .97 | .82 |
| | Attitudes towards the church | M + F | 26 | 18–19 | .38** | .97 | .82 |
| Nelson (1956) | Attitudes towards the church | M + F | 887 | 19–33 | .38** | .76 | .68 |
| | Attitudes towards God as a reality | M + F | 863 | 19–33 | .22** | .76 | .68 |
| Farnsworth (1937) | Attitudes towards war | M | 55 | 18–19 | .30** | .97 | .85 |
| | | M | 50 | 18–20 | .27** | .95 | .85 |
| | | M | 50 | 18–22 | .12** | .91 | .85 |
| Nelson (1954) | Lentz conservatism-radicalism opionnaire | M + F | 901 | 19–33 | .57** | .76 | .70 |

* Significant at .05 level.
** Significant at .01 level.

objects of remote connection to self may be least stable, or subject to greatest change, on this continuum. Next in line of instability might be "opinions" resting on misinformation or unfamiliarity with facts. "Opinions" in the form of superstitions may come third in instability. "Opinions" in the form of deep-seated prejudices would be slightly more difficult to dislodge. "Opinions" founded on early home or religious training and evolving into ethical, or moral, or systematic stereotypes, might be still more difficult to change. Finally, at the extremely stable end of the scale, might be found opinions toward self as intimate self-evaluations—now defined as "adjustments" or "personality traits" or "characteristics." These may be most difficult to modify. Such a continuum of stability would describe in part the differing magnitudes of test-retest correlations found in this study.

We suspect also that part of the explanation may be found in the changes in the society with regard to the particular attitude objects. A view or opinion that is widely prevalent at one time may be less popular at another time. Also, a view that expressed one value at one point in time may represent quite a different value at another time. Thus a view about labor in the 1930's may represent support for the underdog or a humanitarian value, while the same view about labor thirty years later may no longer represent support for the same value. So also views about war may shift with the times to represent quite different values.

Thus we do believe that basic values are likely to remain stable, while their representation in particular attitudes, views, and opinions may shift considerably over time. We believe that a questionnaire getting at a great variety of opinions might be summarized to express a particular basic value at one time, and at another time a somewhat different key or summarization might be determined that would represent the same basic value. We would expect a high correlation (at least one approaching our estimated values) between the two scores, even though the particular opinions might be quite different at the two times. It would be interesting to check this hypothesis against some of the data already secured in the longitudinal studies summarized in Table 5.8.

## Personality

The personality self-report instruments in the field attempt to measure the degree to which an individual possesses a particular characteristic trait. Such a trait approach does not reveal the total personality configuration since it does measure one trait or characteristic at a time. The dynamic interrelation among the characteristics is missing. The point to be made here is that the stability of a particular

Table 5.9. Longitudinal Studies of Personality Traits

| Author and Date | Characteristics Measured or Instruments Used | Sex | N | Ages | Observed — A | Theoretically Expected on the Basis of an Age Curve of Development — B: When Test Reliability Is Perfect | C: Reduced by Actual Test Reliability |
|---|---|---|---|---|---|---|---|
| Eckert (1940) | Bernreuter-Neuroticism | M + F | 50 | 18–19 | .76 | .97 | .78 |
| | Bernreuter-Dominance | M + F | 50 | 18–19 | .80 | .97 | .76 |
| Kelly (1955) | Bernreuter Self-Confidence | M + F | 368 | 25–45 | .61 | .75 | .65 |
| | | M + F | 368 | 25–45 | .47** | .75 | .59 |
| Farnsworth (1938) | Bernreuter-Neurotic Tendencies | M | 55 | 18–19 | .76 | .97 | .84 |
| | | M | 53 | 18–20 | .60 | .95 | .83 |
| | | M | 50 | 18–21 | .69 | .93 | .82 |
| | Bernreuter-Dominance | M | 55 | 18–19 | .73 | .97 | .81 |
| | | M | 53 | 18–20 | .73 | .95 | .80 |
| | | M | 50 | 18–21 | .72 | .93 | .78 |
| Tyler (1961) | Omnibus Personality Inventory | | | | | | |
| | Schizophrenia | M | 402 | 17–21 | .61** | .90 | .75 |
| | | F | 184 | 17–21 | .59** | .90 | .72 |
| | Social Introversion | M | 402 | 17–21 | .67* | .90 | .73 |
| | | F | 184 | 17–21 | .63 | .90 | .69 |

| | Sex | N | Age | r | | |
|---|---|---|---|---|---|---|
| Thinking Introversion | M | 402 | 17–21 | .57** | .90 | .74 |
| | F | 184 | 17–21 | .62** | .90 | .74 |
| Impulse Expression | M | 402 | 17–21 | .62*** | .90 | .70 |
| | F | 184 | 17–21 | .68 | .90 | .69 |
| Taylor (1953) Manifest Anxiety Scale | M + F | 50 | 20–21 | .81 | .97 | .85 |
| Crook (1941) Thurstone Personality Scale | F | 18 | 18–20 | .43** | .95 | .87 |
| | F | 40 | 18–21 | .65*** | .93 | .86 |
| | F | 50 | 18–22 | .48*** | .91 | .84 |
| | F | 60 | 18–24 | .57*** | .87 | .80 |
| Darley et al. (1951) Minnesota Personality Schedule | | | | | | |
| Morale | F | 104 | 19–20 | .70* | .97 | .81 |
| Social Adjustment | F | 104 | 19–20 | .82 | .97 | .82 |
| Schofield (1953) MMPI | | | | | | |
| K Score | M | 83 | 18–20 | .43** | .95 | .87 |
| Depression | M | 83 | 18–20 | .51** | .95 | .80 |
| Psychopathic Deviate | M | 83 | 18–20 | .36** | .95 | .80 |
| Paranoia | M | 83 | 18–20 | .51** | .95 | .77 |
| Psychasthenia | M | 83 | 18–20 | .41** | .95 | .87 |
| Schizophrenia | M | 83 | 18–20 | .38** | .95 | .78 |
| Masculinity-Feminity | M | 83 | 18–20 | .32** | .95 | .75 |
| Plant (1958a) Ethnocentrism Scale | M | 137 | 18–22 | .56** | .91 | .77 |
| | F | 134 | 18–22 | .54** | .91 | .77 |
| Plant (1958b) Ethnocentrism Scale | M | 111 | 18–20† | .66** | .95 | .81 |
| | M | 111 | 18–20‡ | .53*** | .95 | .81 |
| | F | 107 | 18–20† | .67*** | .95 | .81 |
| | F | 107 | 18–20‡ | .57*** | .95 | .81 |

* Significant at .05 level.    † = College group.
** Significant at .01 level.    ‡ = Matched noncollege group.

personality characteristic is an indication of whether a characteristic remains fixed in relation to the normative data rather than whether the characteristic maintains its place in the constellation of personality characteristics for the individual.

The major longitudinal studies on personality questionnaires after age 18 are those within the college years on the Bernreuter Personality Inventory and the Vassar tests.   The Kelly (1955) study also includes Bernreuter Personality Inventory Data over a 20 year period.

We have summarized some of the longitudinal data on self-report personality questionnaires in Table 5.9.   In most cases, the correlations reported are significantly below the estimated values based on an age curve of development.   The major exceptions are usually the studies of one-year duration or those involving the Bernreuter Personality Inventory.

In seeking an explanation for the relatively low level of stability for the self-report personality instruments, we are struck by the susceptibility of these instruments to conscious as well as unconscious distortion by the examinee.   We are led to the conclusion that it is unlikely that the stability of personality can be determined with any degree of precision by these instruments.

## SUMMARY

The longitudinal data available on interests, attitudes and values, and other personality characteristics are far from satisfactory.   A great variety of instruments has been used in the longitudinal studies reported in the literature and only rarely have the data been collected over a period of ten or more years.

We have divided the longitudinal studies in this chapter with respect to type of measuring techniques used, as well as with respect to characteristics being investigated—interests, attitudes and values, and personality characteristics.   In each instance we have compared the correlations reported in the studies with estimates derived from an age curve of development.

In the period from birth to late adolescence, the major types of longitudinal data are based on ratings and observations of the individual by others.   Although this type of evidence is primarily concerned with more surface aspects of personality—the person as seen by others—there is strong evidence of levels of stability which are as high or higher than those reported in the earlier chapters on physical characteristics, intelligence, aptitudes, and school achievement.

We have deliberately selected characteristics for which the levels of stability are approximately the same for girls and for boys, as well as characteristics for which the levels of stability are quite different for the two sexes. Since the evidence is primarily observational, it is not clear whether the difference in the stability values for the two sexes is an artifact of the difficulties of observing and rating males versus females on these characteristics, or whether the differences are attributable to substantial changes in these characteristics as the result of the role expectations and cultural values for the two sexes. It is to be hoped that future research will clarify this issue.

We are unable to apply the Overlap Hypothesis in relation to an absolute scale of personality development to these data because, at present, the notion of an absolute scale of development in this area is not very meaningful. However, we are able to determine how the stability of characteristics, interpreted in terms of consistency of rank order and predicability of the variance on a characteristic from one age period to another, compares with estimates based on an age curve of development.

From these comparisons, it is possible to determine whether the stability of selected personality characteristics in the early years is in agreement with personality theories which suggest rapid early development or whether they are more in agreement with a simple linear development approximating that of an age curve. For the characteristics selected there is ample evidence that for at least one of the sex groups there is far more development in the early years (0 to 5) than would be anticipated on the basis of age curve of development. It is the view of the writer that for these characteristics the early development is, at least quantitatively, in accord with the psychoanalytic literature which suggests that major development of personality takes place in the early years. By an average age of about 2, it seems evident that at least one-third of the variance at adolescence on intellectual interest, dependency, and aggression is predictable. By about age 5, as much as one-half of the variance at adolescence is predictable for these characteristics. Thus, with relatively simple and observable characteristics, in spite of the difficulties in securing useful observational data, there is evidence that the results of longitudinal data are in essential harmony with the theoretical literature on personality development in the early years. Although this statement is subject to many qualifications and limitations, it is clear that personality, at least in the early years, is not a simple development over time. There are periods of very rapid change in this area of human characteristics just as there are in other areas.

It was hoped that more objective and precise evidence bearing on the development of personality in the early years might be found in longitudinal studies using projective techniques and other measuring instruments in which the examinee is likely to have little conscious control over his responses. Although a number of studies do report the *collection* of such data, very little has been done to analyze the data for stability of particular characteristics. Here again it is to be hoped that further research and data analysis will determine the extent to which early development is in line with theoretical formulations.

The instruments which attempt to measure personality characteristics on the basis of cognitive-perceptual tasks do reveal at least as much stability in the period ages 10 to 21 as would be anticipated on the basis of an age curve of development. This would suggest that a considerable amount of change does take place in certain personality characteristics during this period and that any notions of complete personality development by ages 9 and 10 are not consistent with the longitudinal evidence we have found.

In the remainder of this chapter, we have summarized the longitudinal evidence on interests, attitudes, and specific personality characteristics. All this evidence suggests that there is change in these characteristics throughout life and that this change (or stability) is very close to estimates based on an age curve of development. Thus the picture we have of height remaining essentially the same from about ages 20 to 50 is very much different from the picture that emerges from the longitudinal evidence on interests, attitudes, and personality.

However, it should be noted that the age curve is based on the assumption of smaller and smaller amounts of change taking place each year as the base age increases. Thus the change assumed on this basis from ages 5 to 10 is very great in contrast to the change assumed from ages 20 to 25 or 45 to 50.

The self-report instruments used to appraise changes in interests, attitudes, and personality characteristics were intended primarily for use with high school (juniors and seniors) and college students. Thus they are not very useful for determining change in the period of ages 14 to 18 and they are probably not entirely appropriate for the post college years. For the most part, the longitudinal evidence does suggest that more change takes place on these instruments in the first two years of college than in the remaining college years. Furthermore, more change appears to take place in these two years than in the next 10 to 20 years.

We have speculated about the possible explanations for the relatively low levels of stability for the self-report instruments in the age period 14 to 18 and in the age period 22 and up. It is possible that future research, making use of instruments more appropriate to the purpose, will reveal far higher levels of stability than might be expected on the basis of a simple age curve of development. The present self-report instruments do appear to have language and content more appropriate for some age periods than for others and they do not appear to be constructed with great concern for test items which are likely to be stable over time. That is, the questionnaires include a great variety of items, some of which are likely to be answered in a similar fashion over long periods of time, whereas other items are likely to be answered differently as societal and other conditions make for differing interpretations of the questions or statements used.

In this chapter it is quite possible that we have raised more questions for further research than we have answered on the basis of the available longitudinal evidence.

REFERENCES

Adams, F. J., 1957. A study of the stability of broad vocational interests at the high school level. Unpublished Ed. D. Dissertation, New York Univ.

Ausubel, D. P. et al., 1952. A preliminary study of developmental trends in socio-empathy. *Child Develpm.* **23**, 111–128.

Blanchard, B. E., 1950. Recent investigations of social learning. *J. Ed. Res.*, **43**, 50–115.

Bonney, M. E., 1943. The relative stability of social, intellectual, and academic status in grades II to IV and the inter-relationships between these various forms of growth. *J. Ed. Psychol.*, **34**, 88–102.

Bonney, M. E., 1943. The constancy of sociometric scores and their relationships to teacher judgments of social success and to personality self-ratings. *Sociometry*, **6**, 409–424.

Bordin, E. S., and Wilson, E. W., 1953. Change of interest as a function of shift in curriculum. *Ed. and Psychol. Meas.*, **13**, 297–307.

Burnham, P. S., 1942. Stability of interests. *School and Soc.*, **55**, 332–35.

Burgemeister, B. B., 1940. The permanence of interests of women college students. *Archives of Psychol.*, **36**, No. 255.

Cannon, K. L., 1958. Stability of sociometric scores of high school students. *J. Ed. Res.*, **52**, 43–48.

Channing, L., Taylor, K. V. F., and Carter, H. D., 1941. Permanence of vocational interests of high school boys. *J. Ed. Res.*, **32**, 481–494.

Corey, S. M., 1940. Changes in the opinions of female students after one year at a university. *J. Soc. Psychol.*, **11**, 341–351.

Crook, M. N., 1941. Retest correlations in neuroticism. *J. Gen. Psychol.*, **24**, 173–182.

Darley, J. G., 1938. Changes in measured attitudes and adjustment. *J. Soc. Psychol.*, **9**, 189–199.

Darley, J. G., Gross, N., and Martin, W. E., 1951.   Studies of group behavior: stability, change, and contributions of psychometric and sociometric variables.   *J. Abn. and Soc. Psychol.*, **46**, 565–76.

Eckert, Ralph G., 1940.   A mental hygiene approach to speech instruction as a means to personal adjustment.   Unpublished Ph.D. Dissertation, Univ. of California.

Farnsworth, P. R., 1937.   Changes in attitudes toward men during college years.   *J. Soc. Psychol.*, **8**, 274–279.

Farnsworth, P. R., 1938.   A genetic study of the Bernreuter Personality Inventory.   *J. Genetic Psychol.*, **52**, 3–13.

Forer, B. R., 1955.   The stability of Kuder scores in a disabled population.   *Ed. and Psychol. Meas.*, **15**, 166–169.

Getzels, J. W., and Jackson, P. W., 1963.   "The teacher's personality and characteristics," in Gage, N. W. (Ed.), Handbook of research on teaching.   Chicago: Rand McNally and Company.

Gronlund, N., 1959.   Sociometry in the classroom.   New York: Harper.

Grossman, B., and Wrighter, J., 1948.   The relationship between selection-rejection and intelligence, social status, and personality amongst sixth grade children.   *Sociometry*, **11**, 346–355.

Herzberg, F., Bouton, A., and Steiner, B. J., 1954.   Studies of the stability of the Kuder Preference Record.   *Ed. and Psychol. Meas.*, **14**, 90–100.

Herzberg, F., and Bouton, A., 1954.   A further study of the stability of the Kuder Preference Record.   *Ed. and Psychol. Meas.*, **14**, 326–331.

Hudson, M., 1954.   A sociometric study of emotional expression in an elementary school.   Unpublished Ph.D. Dissertation, Univ. of California.

Jacobs, R., 1949.   Stability of interests at the secondary school level.   *Educational Records Bulletin*, **52**, 83–87.

Jersild, A., and Markey, F. V., 1935.   Conflicts between preschool children.   *Child Develpm. Monogr.*, **21**, Teachers College, Columbia Univ., New York.

Kagan, J. and Moss, H., 1962.   From birth to maturity.   New York: Wiley.

Kelly, E. L., 1955.   Consistency of the adult personality.   *Amer. Psychol.*, **10**, 659–681.

Laughlin, F., 1954.   The peer status of sixth and seventh grade children.   Bureau of Publ., Teachers College, Columbia Univ., New York.

Long, L. and Perry, J. A., 1953.   Academic achievement in engineering related to selection procedure and interest.   *J. Appl. Psychol.*, **37**, 468–71.

McKinnon, K. N., 1942.   Consistency and change in behavior manifestations.   Child Develpm. Monogr., No. 30: Teachers College, Columbia Univ., New York.

McNamara, W. J. and Darley, J. G., 1938.   A factor analysis of test-retest performance on attitude and adjustment tests.   *J. Ed. Psychol.*, **29**, 652–664.

Macfarlane, J. W., Allen, L., and Honzik, M. P., 1954.   A developmental study of the behavior problems of normal children between twenty-one months and fourteen years.   Univ. of California Studies in Child Development, Vol. II.   Berkeley: Univ. of California Press.

Moreno, J. L., 1934.   Who shall survive.   Washington: Nervous and Mental Disease Publishing Company.

Morris, D. P., Soroker, E., and Burress, G., 1954.   Follow-up studies of shy, withdrawn children.   I. Evaluation of later adjustment.   *Amer. J. Orthopyschiatry*, **24** 743–754.

Mostovoy, J. L.   Sibling resemblance in intelligence and achievement as related to parental education.   Master's Thesis in progress, Univ. of Chicago.

Nelson, E. N. P., 1954.   Persistence of attitudes of college students fourteen years later. *Psychol. Monogr.*, No. 373.

Nelson, E. N. P., 1956.   Patterns of religious attitudes shift from college to 14 years later. *Psychol. Monogr.*, No. 424.

Paterson, C. H., 1943.   The Vineland Social Maturity scale and some of the correlations. *J. Genetic Psychol.*, **62**, 275–287.

Peck, R. F., and Havighurst, R. J., 1960.   The psychology of character development. New York: Wiley.

Plant, W. T., 1958a.   Changes in ethnocentrism associated with a four-year college education. *J. Ed. Psychol.*, **49**, 162–165.

Plant, W. T., 1958b.   Changes in ethnocentricism associated with a two-year college experience. *J. Genetic Psychol.*, **92**, 189–97.

Powers, M. K., 1956.   Permanence of measured vocational interests of adult males. *J. Appl. Psychol.*, **40**, 69–72.

Reid, J. W., 1951.   Stability of measured Kuder interests in young adults. *J. Educ. Res.*, **45**, 307–12.

Roberts, K. L., and Ball, R. S., 1938.   A study of personality in young children by means of a series of rating scales. *J. Genetic Psychol.*, **52**, 79–149.

Roe, A., 1957.   "Early differentiation of interests," in University of Utah Research Conference on the Identification of Creative Talent, Brighton, Utah.

Rosenberg, N., 1953.   Stability and maturation of Kuder interest patterns during high school. *Ed. and Psychol. Meas.*, **13**, 449–452.

Sanford, R. N. (Ed.), 1962.   The American College. New York: Wiley.

Schofield, W., 1953.   A study of medical students with the MMPl, II.   Group and individual changes after two years. *J. Psychol.*, **36**, 137–42.

Staker, A. M., 1948.   Changes in social status of elementary school pupils. *Ed. Res. Bull.*, **27**, 157–9.

Stoops, J. A., 1953.   Stability of the measured interests of high school pupils between grades 9 and 11. *Ed. Outlook*, **27**, 116–118.

Stordahl, K. E., 1954.   Permanence of Strong Vocational Interest Blanks scores. *J. Appl. Psychol.*, **38**, 423–27.

Strong, E. K., Jr., 1934.   Permanence of vocational interests. *J. Ed. Psychol.*, **25**, 336–344.

Strong, E. K., Jr., 1951.   Permanence of interest scores over 22 years. *J. Appl. Psychol.*, **35**, 89–91.

Strong, E. K., Jr., 1952.   Nineteen year follow-up of engineer interests. *J. Appl. Psychol.*, **36**, 65–74.

Taylor, J. A., 1953.   A personality scale of manifest anxiety. *J. Abn. and Soc. Psychol.*, **48**, 285–290.

Taylor, K. Van F., 1942.   The reliability and permanence of vocational interests of adolescents. *J. Exp. Ed.*, **11**, 81–87.

Thompson, G. G., and Powell, M., 1951.   An investigation of the rating-scale approach to the measurement of social status. *Educ. and Psychol. Meas.*, **11**, 440–45.

Todd, J. E., 1941.   Social norms and the behavior of college students.   Teachers College, Bureau of Publications.

Towner, L. W., 1948.   The relationship between experiences and changes in vocational interests.   Unpublished Ph.D. Dissertation, Univ. of California.

Tuddenham, R., 1959.   The constancy of personality ratings over two decades. *Genetic Psychol. Monogr.*, **60**, 3–29.

Tyler, F. T., 1961.   Stability and variability of various personality measures during the college years.   Paper presented at the American Psychological Association Convention.

Van Dusen, A. C., 1940.   Permanence of vocational interests.   *J. Ed. Psychol.*, **31**, 401–424.

Vernon, P. E., and Allport, G. W., 1931.   A test for personal values.   *J. Abn. and Soc. Psychol.*, **26**, 231–248.

Webster, H. et al., 1962.   "Personality changes in college students," in Sanford, N. (Ed.), The American College.   New York: Wiley, pp. 811–846.

Wertheimer, R. R., 1957.   Consistency of sociometric status position in male and female high school students.   *J. Ed. Psychol.*, **48**, 385–390.

Whistler, L., 1940.   Changes in attitudes toward social issues accompanying a one year freshman social science course.   *J. Psychol.*, **10**, 387–396.

Whitely, R. C., 1938.   The constancy of personal values.   *J. Abn. and Soc. Psychol.*, **33**, 405–408.

Witkin, H. A. et al., 1962.   Psychological differentiation.   New York: Wiley.

Wright, J. C., and Scarborough, B. B., 1958.   Relationship of the interests of college freshmen to their interests as sophomores and as seniors.   *Ed. and Psychol. Meas.*, **18**, 153–158.

# Chapter Six

# ENVIRONMENT

## INTRODUCTION

There is little that is new in the recognition that individuals live in and interact with their environment. No theory of psychology, learning, or growth has ever dismissed the environment as unimportant or to be ignored in accounting for development. Although psychology has always had a place for the environment in its theories, it has not had a corresponding emphasis on the environment in its research procedures, techniques of measurement, or even in its efforts to bring about change in the individual or group.

This is not to advocate a revival of the old nature-nurture conflict which consumed so much emotion, research energy, and publication space in the period 1925 to 1940 in the attempt to give a quantitative determination to the question of the relative influence of heredity and environment on intelligence. This seems to be a sleeping dog which had better be left in a somnolent state. What is advocated is an increased interpenetration of the behavioral sciences for the purpose of making more explicit the causal connections between man's development and the environments with which he interacts. The world with its multiplicity of environments must now be the behavioral scientist's laboratory. Experimental conditions which are impossible to secure in one country because of moral and ethical codes are found to be man's existing conditions in other parts of the world. Child-rearing conditions in one part of the world are in striking contrast with child-rearing conditions in other parts of the world (Whiting and Child, 1953). Educational conditions in one part of the world are in sharp contrast with educational conditions in other parts of the world (Moehlman and Roucek, 1951). The similarities and differences in selected aspects of environments throughout the world now offer opportunities for clarifying many theoretical and empirical issues in our understanding of man and his development.

As the means of communication among behavioral scientists throughout the world improve to the level already attained in the physical and biological sciences, we may secure more satisfactory answers to the question Anastasi (1958) used as the title of her paper, "Heredity, environment and the question How?" Although we would wish to emphasize the use of modern testing and sampling theory in this quest, it is clear that the basic questions have to do with "how" rather than with "how much."

What is to be gained by reviving or reemphasizing an environmental point of view? Are there ways of using an environmental point of view to add to our ability to understand and predict human growth and development?

Perhaps a brief anecdote is in order at this point. The writer spent several months attempting to analyze the data on growth in height reported by Tuddenham and Snyder (1954). The correlation between height at ages 3 and 18 for males was +.76 as reported in this work. For many fruitless weeks the writer attempted all sorts of combinations of earlier height measurements, gain measurements, ratios, etc., but try as he would, the correlations between height measurements at ages 3 and 18 never rose above +.80. After considerable frustration, the writer toyed with the notion that weight related to height might improve the prediction, but here too most efforts did little to change the relation between height at age 3 and age 18.

Turning to the Dearborn et al. (1938) data from the Harvard Growth Study, we found it possible to use the extremes of socio-economic status as a crude index of the environment as it related to the nutritional care as well as health support of the individual. Using this index plus height at age 7, the multiple correlation with height at age 17.5 rose to +.90 in contrast with the correlation of +.74 reported by Shuttleworth (1939) for all groups combined for these ages.

The point to be emphasized is that the *introduction of the environment as a variable makes a major difference in our ability to predict the mature status of a human characteristic.* Thus there is empirical as well as theoretical support for the use of the environment in our attempt to explain and predict growth and development. In this chapter we shall try to indicate the extent to which this generalization applies to different characteristics as well as the considerations and refinements necessary if it is to be put to a critical test.

MEASUREMENT OF INDIVIDUAL VERSUS
MEASUREMENT OF ENVIRONMENT

The pioneers in the field of individual differences such as Wundt, Galton, Cattell, and Binet were primarily interested in describing the ways in which individuals varied with respect to a particular characteristic or behavior. They were able to demonstrate that when a particular measure such as reaction time, discrimination of sounds, intellectual performance, etc., was used, individuals varied greatly.

In their early work, these men wished to measure individual differences without taking into consideration the differences in the backgrounds and experiences of the persons being measured. Binet and Simon (1905) preferred that nothing be known by the tester about the individual prior to the testing. They wished to regard the examinee as "an X to be solved" by determining what he could do under standard testing conditions on a standard set of tasks. Although Binet and Simon did recognize that the test results were not entirely accurate if the individual had grown up under deprived conditions, for them this was but one source of error in the testing procedures which might be corrected by appropriate weighting of the results or through use of only selected portions of the test.

Thus the emphasis in the testing was on individual differences with little attention given to environmental variation except as a potential source of error in the measurement of individuals. This early emphasis in the testing movement has continued to the present.* Our catalog of tests of individual differences is enormous, whereas our instruments for measuring environmental differences consist of a few techniques for measuring social class status and socio-economic status.

SPECIFICITY OF THE ENVIRONMENT

Although it is true that we have had a few environmental measures (social class, etc.), these have been relatively crude measurements of

* It is only fair to note that some of the major longitudinal studies such as the Fels Study and the Berkeley Growth Study (Baldwin et al., 1949; Macfarlane, 1938; Sontag, et al., 1953; and Kagan and Moss, 1962) have attempted to rate the home environment on a number of characteristics and have related these ratings of the home environment to individual differences. More recently an instrument has been devised by Pace and Stern (1958) to measure college characteristics. It is also true that social workers, sociologists, and anthropologists have given a great deal of attention to descriptions of the immediate environment as well as the larger milieu. However, individual differences and environmental variations are not usually systematically related.

environments on a scale of general merit.   For example, it has been assumed that an index of social status and economic well-being describes a continuum which is meaningful and operational.   It would be idle to contend that this has not been useful, since it has been of considerable predictive value in many areas.

However, just as a general index of intelligence or I.Q. has obscured many of the very important differences among individuals, so the general index of social or economic status has obscured many very important differences among environments (Kahl, 1953).   The research we have attempted to describe in the preceding chapters has indicated a few of the dimensions along which environments may vary.   We find great need for measurements of a number of different environmental variables which, we suspect, may not be very highly related to each other.   Two environments which may be equal in their effect on physical growth may be quite different in their effects on intellectual growth or school achievement and may be even more variable in their effects on attitudinal and emotional development.

It is likely that factorial research which has proven so powerful in the identification of the major dimensions on which individuals differ may prove to be equally powerful in defining the dimensions on which environments differ.

We are led by the available evidence to begin by regarding environments as having a number of highly specific characteristics and as a result having highly specific consequences for human growth and development.   We do not doubt a proposition that two environments similar in many characteristics and different in a few may have markedly different effects on the individuals who are in them.   That is, the whole is likely to be more than a simple summation of the parts. We do suggest that the strategy of research on environmental variation begin with the attempt to describe and measure the specific characteristics of environments and then proceed to the study of the consequences of various combinations of these specific characteristics.

Since we began this study with an attempt to describe human characteristics and their stability and change, we are led to the search for the environmental conditions which alter or stabilize these individual characteristics.   From this point of view, we would suggest that the initial attempts to describe environmental conditions be related to the individual characteristics already identified.   Initially, the description and measurement of environmental differences would be in the same terms as those already in use for the description and measurement of individual differences.   Undoubtedly, future research may reveal a classification and measurement scheme for environments which will differ fundamentally from the scheme now found to be useful in research on individual differences. Although this may be true, our major plea is that until individual and environment are described in the same or at least in congruent terms, the task of determining consequences and effects cannot proceed with much power or precision. The Murray (1938) concept of needs and press represents a step in this direction as does the concept of role (Sarbin, 1954; Stern, Stein, Bloom, 1956) as an intermediary term between individual and environment.

## MEANING OF ENVIRONMENT

By environment, we mean the conditions, forces, and external stimuli which impinge upon the individual. These may be physical, social, as well as intellectual forces and conditions. We conceive of a range of environments from the most immediate social interactions to the more remote cultural and institutional forces. We regard the environment as providing a network of forces and factors which surround, engulf, and play on the individual. Although some individuals may resist this network, it will only be the extreme and rare individuals who can completely avoid or escape from these forces. The environment is a shaping and reinforcing force which acts on the individual.

The term *environment* must be understood and described in such a way that regularities and generalities may be recognized. The environment as the totality of forces affecting the individual is so complex as to be impossible to handle by present research methods. At the level of total environment, each individual may be said to have lived in a unique environment and no two individuals have had the same combination of environmental factors. If we think of environments as giving opportunities for interaction and experience, it may be contended that no two individuals have had the same interactions and experiences. Furthermore, if we conceive of major change in the individual as arising out of a single experience, much of our research would reduce to the psychoanalytic interview searching for the powerful experiences in each individual's life.

Somewhere between the total environment and the specific interaction or single experience is the view of the environment as a set of persisting forces which affect a particular human characteristic. Such a view of the environment reduces it for analytical purposes to those aspects of the environment which are related to a particular characteristic or set of characteristics. We may, since we are on the outside of the individual, conceive of the environment in terms of the probability that it provides for selected experiences or interactions. Thus an environment which has a higher probability of providing certain experiences than another environment may be said to be a more powerful environment insofar as the appropriate human characteristic is concerned.   This view of the environment would be in general agreement with the *Alpha Press* (the environment as viewed by observers) described by Murray (1938).   In contrast, Murray's *Beta Press* (the subject's perception of the environment) would not be included in this attempt to describe the environment in operational terms.

Furthermore, we are most interested in the environment over time. The environment at a particular moment is of interest only as one segment in a larger time sampling of the environment.

We believe it to be possible to find ways of describing and measuring environments as they affect and, to some extent, determine particular human characteristics.   It is likely that our knowledge in a number of areas is already sufficiently great to permit the development of such environmental measures.   Thus we may describe with considerable precision the environment that affects stature and physical growth. Nutritionists, physical anthropologists, and medical specialists are able to specify some of the major characteristics in an environment which will positively or negatively affect height growth.   These include nutritional elements as well as physical and medical care.   However, the environmental measure must include more than the availability of a supply of such elements and care; it must also determine the extent to which they are utilized by the individuals under study.

Psychologists, educators, and human development specialists are able to specify some of the major characteristics of an environment which will positively or negatively affect the development of general intelligence or school achievement.   These include communication and interaction with adults, motivation and incentives for achievement and understanding of the environment, and the availability of adult models and exemplars of language, communication, and reasoning.   Here again, it is not only the availability of the elements but also the extent to which the individual interacts with and makes use of these elements.   A major attack on this problem has been made by Wolf (1963) and Dave (1963).*

SOME ENVIRONMENTAL VARIABLES

We often tend to think of environments in such terms as good or poor, desirable or undesirable, wholesome or unwholesome, etc.   In part,

* See Chapter 3, page 78 and Chapter 4, page 124.

this inclination may be attributed to the very small number of environmental measures available and to the general tendency to think of wealth, high social position, and professional occupational status as being indices of good environments, whereas poverty, lower social position, and unskilled occupational status are regarded as indices of poor environments. Although it is undoubtedly true that wealth favors the individual in many ways, it is quite possible that lack of wealth may facilitate the development of certain characteristics. It is unlikely that environments can be classed as good or bad in some total way. Furthermore, the use of evaluative terms may hamper the attempt to secure operational definitions of environments and may hamper our efforts to study the interrelationships between environments and the development of selected behavioral characteristics.

Newman, Freeman, and Holzinger (1937) provide some evidence on the specificity of environments. They studied 19 pairs of identical twins who had been separated during early childhood and measured each twin on a number of characteristics. They also attempted to secure a great deal of descriptive material on the environments in which each of the separated twins lived. Each environment was rated on a scale of 5 to 50 with respect to educational conditions, social conditions, and physical and health conditions. Table 6.1 shows the correlations between the differences found in each pair of children and the differences found in the two environments in which the children were reared. It will be noted that the *physical and health environmental* differences ratings are most clearly related to the differences in the weights of the twins, but have low relations with other characteristics. The *educational environmental* differences are highly related (+.91) to the differences in the tested school achievement of the twins

*Table* 6.1. *Correlations of Twin Differences on Certain Traits with Estimated Differences in Three Environmental Ratings*
(*Adapted from Newman, Freeman, Holzinger,* 1937)

| | Environmental Difference Rating | | |
| | | | |
| *Characteristic* | *Educational* | *Social* | *Physical and Health* |
|---|---|---|---|
| Height | −.015 | −.005 | −.175 |
| Weight | −.095 | .226 | .599 |
| Binet I.Q. | .791 | .507 | .304 |
| Stanford Educational Age | .908 | .349 | .139 |

and are also highly related to differences in Binet I.Q. (+.79).   In turn, the *social environmental* differences are only moderately related to differences in both I.Q. and school achievement.   The point of these comparisons of environmental and twin differences is the specificity of the relationships between environmental and individual characteristics. The dimension of the environment which is highly related to one human characteristic may be unrelated to another characteristic.

From this evidence it would seem that the research strategy which is likely to be most effective will be one in which theory and empirical research are directed toward relating particular environmental measures to particular human characteristics.   Some leads toward such research are provided by the longitudinal and other research we have cited in the preceding chapters.   Some of the environmental factors which are most clearly related to the following human characteristics are likely to be the following.

## Stature

Differences in height are likely to be related to

1. Quantity and quality of nutrition provided by the environment and utilization of appropriate nutritional elements by the individual.

2. Sanitation, disease prevention, and the availability and utilization of medical care when needed.

## General Intelligence

Differences in general intelligence are likely to be related to

1. Stimulation provided in the environment for verbal development.

2. Extent to which affection and reward are related to verbal-reasoning accomplishments.

3. Encouragement of active interaction with problems, exploration of the environment, and the learning of new skills.

## School Achievement

Differences in school achievement are likely to be related to

1. Meaning which education comes to have for one's personal advancement and role in society.

2. Level of education of and value placed on education by the significant adults in the individual's life.

3. Extent to which school achievement is motivated and reinforced by parents or significant adults in the individual's life.

EVIDENCE ON THE EFFECTS OF ENVIRONMENTS

The study of twins and siblings reared together and reared apart yields striking evidence of the effect of the environment on the development of specific human characteristics, especially when the environments as well as the individuals have been measured (Husén, 1959; Burt, 1958; Newman, Freeman, and Holzinger, 1937).*

Less clearcut evidence is provided by cross-sectional studies in which groups of individuals living under one set of environmental conditions are compared with groups which have lived under contrasting conditions. We have referred to such evidence in the preceding chapters.

Perhaps the most direct evidence of the effects of environments is to be found in longitudinal studies where there is not only repeated measurements on the individual but in which there is also evidence on the environmental conditions under which each person has lived during the period of time under study. Throughout this work, we have cited some of the more crucial studies bearing on this point.† Since these studies have been made under a variety of research conditions, comparable data are not available for all the studies.

It will be noted that the correlations between measurements obtained at two different ages are usually much lower when the environment is not considered than when the environment is considered as a third variable. We have hypothesized that the correlations between measurements at any two ages would be theoretically perfect if the relevant environmental conditions were measured accurately and taken into consideration in the correlations. Although some of the correlations for periods as long as ten years do in fact approach the range of +.90 to +1.00, the departure from correlations of unity in these studies does, we believe, arise from the, as yet, crude measurements of the environment. In each case we have utilized approximations or relatively indirect indices of the relevant environmental factors. Thus we have used occupational level as an index of the quality of nutrition provided by the environment—a very indirect and, we believe, poor approximation of the quality of the environment for height growth. Similarly, we have used parent's education and occupation as an index of the quality of the environment for intelligence and scholastic achievement—a relevant but poor approximation.

However, even with such crude environmental indices, the improvement in the correlations is substantial. The correlations reported on

* See Chapter 2, Table 2.3; Chapter 3, Tables 3.4, 3.4a,; Chapter 4, Tables 4.3, 4.4.
† See Chapter 3, Charts 6, 7; Chapter 4, Charts 4, 5.

longitudinal measurements when environmental measures are also included do suggest that further improvements in environmental measures are likely to result in multiple correlations approaching unity.

What is also clear in these studies is the relatively high correlations between the gain measures and the environmental conditions in contrast to the approximately zero relation between the initial and the gain measures.

We believe that three major propositions can now be made about longitudinal measurements:

1. The correlation between measurements on the same characteristic at two different times is a function of the Overlap Hypothesis when the environment in which the subjects have lived during the intervening period is *not* known or considered.

2. The correlation between measurements on the same characteristic at two different times will approach unity when the environment in which the individuals have lived during the intervening period is known and taken into consideration.

3. The gains made by individuals subjected to the same powerful environment will tend to be equal.

These three propositions are most clearly seen in the selected data on school achievement, and intelligence.*

In each case it may be seen that the correlations and consistency indices are much lower when the environment is ignored than when the two contrasting environments are differentiated. Although we rarely find all individuals making identical gains, it will be seen that the majority of individuals within each environment have similar slopes in the lines connecting the initial and terminal measurements while the slopes of the lines are very different between the two contrasting environments.

As we study these charts, we will observe the increasing differentiation of individuals growing up under different environmental conditions. It is apparent in these charts that much of what we usually describe under the heading of individual differences might more accurately be termed environmental differences. Two individuals who had identical measurements on a specific characteristic at one time will at a later time (if they have lived under very different environmental conditions) have very different measurements. Put in other terms, individuals growing up under one set of environmental conditions may be disadvantaged in the development of a specific characteristic in contrast to individuals growing up under more favorable conditions.

* See Chapter 3, Charts 6 and 7 and Chapter 4, Charts 4 and 5.

Such differences in environments in the periods of most rapid growth may make substantial differences in the career and history of individuals. We are not able to state glibly what is good or bad for the individual or for the society. However, we are able to state emphatically that the conditions under which individuals live during the period of most rapid development for a particular characteristic will have far reaching consequences for the qualitative and quantitative development of *that* characteristic and that this development, in turn, will have far reaching consequences for each individual's conditions of life, career, and sources of fulfillment and happiness.

The nature of the individual's pursuit of life, liberty, and happiness may be largely determined by the nature of the environmental conditions under which he has lived in his formative years. Furthermore, although individuals in a democracy may not be equal at birth, much of their inequality at maturity may be ascribed to the lack of *equality of opportunity* if we see opportunity and environmental conditions as partial reflections of each other.

## DIFFERENTIAL EFFECTS OF ENVIRONMENTS AT DIFFERENT PERIODS OF GROWTH

Throughout this work, it has been evident that growth in each characteristic is not in equal units per year of development, experience, or learning. There appear to be periods in which growth and development is very rapid and periods in which growth is very slow or in which there is little perceptible change. Characteristically, growth curves are very rapid in the early stages and relatively slow at later stages. In addition, there do appear to be marked changes in some characteristics at the adolescent stage or in the early stages of encounter with a markedly different environment (for example, refeeding after famine or malnutrition, entrance to college, placement in a foster home, etc.). We have considered the magnitude of change at different periods for selected characteristics. Growth in stature is an extreme example where the growth during the nine month period from conception to birth is of the same absolute magnitude as growth during the nine year period from ages 3 to 12.

It appears to us to be a tautological statement that the environment can have no influence on development after the full development of a stable characteristic has been attained. Thus, after full physical growth in stature has been completed by about age 20, we cannot see any way in which the quality of the environment can change stature (except in the case of special drugs or rare diseases). In the period of ages 13 to 20 in which boys attain their last 10% of full stature, the quality of the environment can have some effect on height growth. However, we estimate this effect will be limited to a variation of about 1% of full stature. In contrast, the period from conception to age 2.5 accounts for about 50% of full stature development, and we believe that variations in the quality of the environment at this stage can affect mature stature by as much as about 5%.

Beginning with the hypothesis that variation in the environment may affect height at maturity by as much as 10%, we would expect the effects of the environment to be greatest during the period of most rapid growth and least during the period of slowest growth. In the various chapters we have attempted to relate the magnitude of growth in each characteristic at each stage to the variability in the environments. In general, the data available do support the generalization that the effects of marked differences in the environment are greatest in the period of greatest normal growth and least in the periods of least normal growth. Thus variations in the environment have significance for individuals in direct proportion to the rate of normal growth during the particular time periods involved. Furthermore, a shift from one environment to another will have greatest consequence in a period of rapid normal growth and will have little effect on the individual during the period of least rapid normal growth.

## ENVIRONMENTAL CONSTANCY AND CONSISTENCY

Our research suggests that although the environment may have its greatest effect on individuals in the first year or so that they are within it, its effect is stabilized and reinforced only when the environment is relatively constant over a period of time. Where the relevant environment is similar for a group of individuals and is constant over a long period of time (such as five years) the correlation between the initial and retest measurements of the individuals in this environment is likely to approach unity. This may be explained on the basis that all individuals may not respond to the environment at the same points in time and the periodicity in individual growth may be "blanketed" by a sufficiently long interaction of the individuals with the environment so that all are affected by it.

Constancy also implies a further stabilization and reinforcement of the change which has taken place. Once the initial alteration has been made, the overall environment further supports and sets it so that it will be resistant to small variations which could take place as the result of the influence of subaspects of the environment.

Closely related to the idea of constancy is that of consistency. Constancy implies a similarity over time, whereas consistency suggests that various contemporary aspects of the environment are similar and mutually reinforcing. A characteristic such as physical growth may be understood in terms of constancy of nutritional elements, the availability of a supply of nutriments over time, and the availability of physical and medical care which is preventive of major diseases or which assures prompt remediation to reduce the ravages of such diseases on growth. However, the more complex types of growth (emotional as well as intellectual) may be affected only when there is considerable consistency in the environment as different individuals and ideas interact with the subjects (or learners).

Perhaps the notion of consistency is what distinguishes a powerful learning environment from one that is only moderate or ineffectual in its consequences for the students. Thus the evidence in Chapter 4 suggests that growth in problem solving is minimal and insignificant if only a single course emphasizes this type of thinking and learning. Whereas, if all portions of the curriculum emphasize and encourage this type of thinking, change is likely to be substantial for almost all the students who complete a year or more of the learning program. Much the same factor appears to be at work in the Bennington program described by Newcomb (1943) in which a socially liberal attitude was stressed in curricular as well as extracurricular activity. As a result of this environment, the Bennington students changed significantly on a liberalism scale over a four year period, whereas students in other colleges where the environment did not consistently emphasize social liberalism did not change significantly over the same period.

## ENVIRONMENTAL EFFECTS VERSUS INDIVIDUAL-ENVIRONMENT TRANSACTION

The types of environments we have been attempting to understand are both very powerful and extreme. Thus we have attempted to identify very abundant and very deprived environments—the extremes of particular continua. These extreme environments appear to have powerful effects, so much so that almost everyone in them is affected in similar ways.

For these extreme environments, the effect is similar for almost all individuals who are in them during the period of time involved. Although a few individuals may, for various reasons, escape partially from the effect of these extreme environments, the proportion of individuals affected in similar ways is likely to be very high, perhaps 90% to 95%. It would seem appropriate to regard these environments as having a one-way effect in that the individual is affected or altered by the environment and he, in turn, is relatively powerless to affect or alter the environment. It would seem far from accurate to describe the relation between the individual and the environment in such extreme instances as interaction or transaction.

Extreme environments that are sustained long enough are likely to have powerful effects on all individuals in them. The effect of these extreme environments is especially dramatic on young children since such children are unable to effect any physical or psychological escape from these overpowering environments. However, the type of effect on young children is not limited to such extreme environments. For the most part, young children may be affected by many environments in this one-way type of effect. That is, the young child is likely to be both plastic and helpless in the face of the environment in which he lives—he knows no other—and he can be directly affected by the pressure and the demands of the environment.

As individuals leave one environment and enter another they seem to be especially susceptible to the effects of the new environment in the initial period in the new environment. Several of the studies reported in this work (Lee, 1951; Newcomb, 1943; Webster et al., 1962; and Dressel and Mayhew, 1954) suggest that changes in the individuals are greater in the first time unit (a semester, year, etc.) in the new environment than in succeeding units of time. Perhaps this is because of the individual's fuller acceptance of the environment and its demands while he is still a "freshman" in it. It is also possible that he is unable in this initial period to discover subenvironments in which he may live with less discomfort or that he is unable to develop a set of defenses against the demands of the new environment until he has been in it for an extended period of time.

We do regard it as possible for a few individuals to be sufficiently powerful to alter or at least effect some modification in the environment. These individuals, it seems to us, are likely to be persons who, finding the environment one which they cannot adapt to or which is uncomfortable for them, are impelled to do something which will make the environment more nearly like one which they have experienced previously. Such individuals may band together and by their pooled efforts bring about changes in the larger environment.

We suspect that the interactive relation between environment and individuals is most likely for older and relatively extreme individuals. Such individuals may have in mind an environment which they have already experienced and to which they did make effective adaptations. Having such a model in mind, they may seek to produce appropriate modifications in their present environment. We would speculate that for a true interactive situation to take place the environment must be somewhat plastic and less powerful than the extreme environments we have been considering. In any case, the individuals must be older and have had experience with other environments as well as be dissatisfied with aspects of the newer environments.

The interactive situation may also be present when individuals are motivated to attempt to create a utopian environment. Such a utopian environment may be created out of a combination of ideas and/or experience with previous environments. Here again we would expect that only a small number of individuals actually do seek to develop utopian environments. Perhaps in the future this may become more common as planning specialists supported by behavioral science research are given political and economic power to modify present environments or even to create new ones.

ENVIRONMENT RESISTANCE AND ENVIRONMENT SELECTION

Search as diligently as we can and analyze the data as carefully as possible, we are unable to find any studies in which the impact of the environment is identical for all individuals in the environment. Although we do come very close to similar gains and growth measures, in no study do we find identical gains for all individuals. We are of the opinion that some of the differences among individuals in the gains may be ascribed to the lack of precision in our measurements of the environments and that we will find a closer approximation to our postulate of equality of growth under identical powerful environments when we improve our environmental measures. However, we do not believe that we are likely to ever attain this full equality even when we have attained maximum validity and precision in environmental measures.

We begin to comprehend this limit more fully as we read Newcomb's (1943) *Personality and Social Change*. Although the changes in social attitudes of the Bennington students over a four year period were considerable and very different from the changes estimated for the students at Skidmore and at Williams College, Newcomb does find some variation in the attitude changes of students at Bennington. Although the external environment at Bennington was very similar for all the girls, it is evident from Newcomb's analysis that some girls were more influenced by this external environment than were others. Newcomb differentiates two types of groups which were least responsive to the effects of this external environment.

The first group consisted of individuals who, because of personal emotional problems, were not very responsive to the environment. Such individuals were apparently not in active communication with the larger environment because of their personal concerns. We can conceive of these as individuals who are largely encapsulated by their own concerns and who live in a world largely of their own making. If individuals are isolated and are in poor contact with the larger environment around them, they cannot be affected as much by that environment as will individuals who are in active interaction and engagement with the external environment.

The second group described by Newcomb consisted of individuals who preferred to live in small subenvironments. Such subenvironments are largely made up of cliques of persons with whom the individual is in greatest sympathy. Such individuals place the larger environment at a remote distance and consequently are less affected by it. In a sense, these individuals are also living in some isolation from the larger environment. They are thus less likely to be motivated and stimulated by the larger environment.

In the research reported by Newcomb, a small proportion of individuals was isolated from the larger environment. We would estimate that the proportion of individuals who are able to effectively resist or ignore the impact of a powerful environment must be quite small.

As we move from extreme—and relatively powerful environments— to less extreme environments we are of the opinion that the individual will have a greater possibility of resisting the effects of the environment. Such an environment is less demanding and places the individual under less pressure.

As we move from relatively constant and consistent environments to more complex environments which have many subenvironments, the individual is given an opportunity to select the environment or environments with which he will interact (for example, a small denominational college in contrast to a large state university). We believe that when given a choice of subenvironments the individual is most likely to select environments which are most in harmony with previously experienced environments. That is, the individual is likely to select an environment which does not threaten the defenses and modes of interaction with the world that he has already developed.

SUMMARY

In this chapter we have emphasized the effect of the environment on both stability and change. Where the environment is relatively constant over long periods of time we have hypothesized that a relevant human characteristic will be far more stable than when the environment is more changeable. We have pointed to a few instances in which the combination of an environmental index with a measurement of an individual characteristic results in markedly higher correlations with the individual characteristic at a later point in time than is demonstrated when no information about the environment is included.

On the other hand, when the environment shifts markedly from one point in time to another, stability is likely to be lower than would be expected from the Overlap Hypothesis. In addition, where the environmental change is an extreme one, groups of individuals exhibit characteristic changes which can be attributed to this shift in environments.

We have further pointed up the need for specification and measurement of particular aspects of the environment as they relate to particular human characteristics. Although our present environmental indices are relatively gross and general, we anticipate that major shifts in our understanding of growth and development will be likely to occur when we develop more adequate and precise environmental measures. Here is an area in great need of new research and instrumentation. From the research already completed, we have sketched in brief outline some of the dimensions that this new research and instrumentation should take in relation to specific areas of human growth and development.

We have stated three major propositions about longitudinal measurement.

1. The correlation between measurements of the same characteristic at two different times is a function of the Overlap Hypothesis when the environment in which the subjects have lived during the intervening period is *not* known or considered.

2. The correlation between measurements of the same characteristic at two different times will approach unity when the environment in which the individuals have lived during the intervening period is known and taken into consideration.

3. The gains made by individuals subjected to the same powerful environment will tend to be equal.

The variation in environments is not likely to have similar effects at all stages in the development of a characteristic. We have suggested that the environment will have its greatest effects in the period of most rapid normal development of a characteristic and that its effects will be least in the period of slowest normal development, while its effects will approach zero in the period of no normal change in the characteristic. On the other hand, at late stages in the normal development of a characteristic, only the most powerful and consistent environments are likely to produce marked changes in either the individual or the group.

Although it is likely that the most frequent relation between individuals and environments is one way in that the individuals are affected by the environments rather than vice versa, there are undoubtedly situations in which individuals may alter the environment or selectively interact with subenvironments. Such situations may occur when the environments are less extreme and less powerful, when the individuals are more mature and/or powerful, or where the individual for some reason is not in active contact with the larger environment.

REFERENCES

Alexander, M., 1961. The relation of environment to intelligence and achievement: a longitudinal study. Unpublished Master's Thesis, Univ. of Chicago.

Anastasi, A., 1958. Heredity, environment and the question "How?" *Psychol. Rev.*, **65**, 197–208.

Baldwin, A. L., Kalhorn, J., Breese, F. H., 1949. The appraisal of parent behavior. *Psychol. Monogr.*, **63**, No. 4, Whole No. 299.

Binet, A., and Simon, T., 1905. Methodes nouvelles pour le diagnostic du niveau intellectual des anormaux. *Annee Psychol.*, **11**, 191–244.

Burt, C., 1958. The inheritance of mental ability. *Amer. Psychologist*, **13**, 1–15.

Dave, R. H., 1963. The identification and measurement of environmental process variables that are related to educational achievement. Unpublished Ph.D. Dissertation, Univ. of Chicago.

Dearborn, W. F., Rothney, J. W. M., and Shuttleworth, F. K., 1938. Data on the growth of public school children. *Monogr. Soc. Res. Child Develpm.*, **3**, No. 1.

Dressel, P. L., and Mayhew, L. B., 1954. General education: explorations in evaluation. Washington: American Council on Education.

Husén, T., 1959. Psychological twin research. Stockholm: Almquist and Wicksell.

Kagan, J., and Moss, H., 1962. From birth to maturity. New York: Wiley.

Kahl, J. A., 1953. Educational and occupational aspirations of 'comman man' boys. *Harvard Educ. Rev.*, **23**, 186–203.

Kirk, S. A., 1958. Early education of the mentally retarded. Urbana, Illinois: Univ. of Illinois Press.

Learned, W. S., and Wood, B. D., 1938. The student and his knowledge. New York: Carnegie Foundation for the Advancement of Teaching, Bulletin No. 29.

Lee, E. S., 1951. Negro intelligence and selective migration. *Amer. Soc. Rev.*, **16**, 227–233.

Macfarlane, J. W., 1938. Studies in guidance. Washington: *Soc. Res. Child Develpm.*, **3**, No. 6.

Moehlman, A., and Roucek, J. S., 1951. Comparative education. New York: Dryden Press.

Murray, H., 1938. Explorations in personality. New York: Oxford Univ. Press.

Newcomb, T. M., 1943. Personality and social change. New York: Dryden Press.

Newman, H. H., Freeman, F. N., and Holzinger, K. J., 1937. Twins: A study of heredity and environment. Chicago: Univ. of Chicago Press.

Pace, C. R., and Stern, G. G., 1958. An approach to the measurement of psychological characteristics of college environments. *J. Educ. Psychol.*, **49**, 269–277.

Sarbin, T., 1954. "Role theory," Chapter 6, in Lindzey, G., Handbook of social psychology, Vol. I. Reading, Mass.: Addison-Wesley, pp. 223–258.

Shuttleworth, F. K., 1939. The physical and mental growth of girls and boys age six to nineteen in relation to age at maximum growth. *Monogr. Soc. Res. Child Develpm.*, **4**, No. 3.

Sontag, L. W., Baker, C. T., Nelson, V. L., 1958. Mental growth and personality development: a longitudinal study. *Monogr. Soc. Res. Child Develpm.*, **23**, 1–143.

Stern, G. G., Stein, M. I., and Bloom, B. S., 1956. Methods in personality assessment. Glencoe, Illinois: Free Press.

Tuddenham, R. D. and Snyder, M. M., 1954. Physical growth of California boys and girls from birth to 18 years. *Child Develpm.*, **1**, No. 2. Berkeley: Univ. of California Press, pp. 183–364.

Webster, H. et al., 1962. "Personality changes in college students," in Sanford, N. (Ed.), The American College. New York: Wiley, pp. 811–846.

Whiting, J. W. M., and Child, I. L., 1953. Child training and personality: a cross-cultural study. New Haven: Yale Univ. Press.

Wolf, R. M., 1963. The identification and measurement of environmental process variables related to intelligence. Ph.D. Dissertation in progress, Univ. of Chicago.

# Chapter Seven

# CONCLUSIONS AND IMPLICATIONS

## THE EARLY STABILIZATION OF SELECTED CHARACTERISTICS

The major evidence used throughout this work is based on longitudinal studies in which the same individuals have been repeatedly measured at different points in their lives. This longitudinal evidence enables us to draw a quantitative picture of the development and growth of selected human characteristics for individuals as well as groups. The group data can be represented in a variety of ways, but the most common procedure used in this book is to show the linear correlations between measurements of a characteristic at different age periods.

One major finding that emerges from these longitudinal studies is that the correlations between the measurements of a *specific characteristic* at particular ages are very similar from one study to another. When allowances are made for the variability of the samples, the comparability of the measuring instruments, and the reliability of the measurements, the results of the different studies are strikingly similar. The results of a longitudinal study reported, for example, in 1921 are in close agreement with the results of a study reported in 1954. The comparability of the correlational pattern from studies done by different workers, with different samples, at different times, and under different conditions has been a constant source of surprise to the writer. Perhaps he has been oversensitive to the idea that in the behavioral sciences, and especially in education, our generalizations must be of a very limited nature in contrast to the more powerful and sweeping generalizations of the natural sciences.

When appropriate allowances are made for the sampling variations and the measurement problems, longitudinal data begin to yield such similar patterns of *relationships* that we can begin to think in terms of *laws* rather than *trends*. A single curve can be used as a very close approximation to the results found in many different studies. Perhaps the major point of this is that the congruence of the many different quantitative findings permits us to draw powerful generalizations in what have hitherto been regarded as the "soft" sciences. The consistency of the data when summarized in terms of relationships gives great promise that investigations of the underlying variables and determinants will yield increased understanding of the ways in which growth and development take place and of the forces which may affect these developments.

One approach to this understanding has been to compare the longitudinal data with the normative or cross-sectional data obtained on different samples. When the normative data are reinterpreted in terms of scales which have equal units and a defined zero, it is possible to estimate the magnitude of the quantitative growth on a particular characteristic from one period of time to another. From these data it is possible to sketch the approximate curve showing the theoretical development of each characteristic over time. The theoretical curve is based on the assumption of perfectly reliable and comparable measurements with the full variability found in the population rather than that to be found in a restricted sample. When these assumptions are met in longitudinal data or when the longitudinal data are treated in such a way as to satisfy these assumptions, there is a high degree of agreement between the curves of development based on longitudinal data and the curves of development based on cross-sectional data. The agreement is so high that one curve can be substituted for the other with little loss in accuracy or detail.

The use of theoretical curves of development has enabled us to summarize many longitudinal studies for each set of characteristics. Such summaries reveal the basic agreement in the findings of many independent studies conducted during the past half century. This agreement is so great that we may expect new studies to be placed within this larger matrix in such a way that the findings can, in large part, be anticipated on the basis of the theoretical curves as well as the prior studies in the field.

From this summarization of longitudinal research we are beginning to understand the basic dimensions of growth and development in so-called stable characteristics. There is a cumulative quality to the many studies in this field such that each new study can be appropriately placed in this growing body of knowledge. Furthermore, where an investigation yields findings not in harmony with the existing body of data, the researcher must now look further into his results to ascertain why they differ from those theoretically expected. In each area, there are some obvious explanations which he should explore further, such as the nature of the measurements, the reliability and accuracy of the findings, and the special characteristics of the sample. If these do not account for the differences, then the researcher may wish to probe further to determine whether there is, in fact, a new set of phenomena being revealed by his investigation. Thus the very nature of research methodology becomes in part altered by the consolidation of previous longitudinal data and by the availability of theoretical as well as observed values.

Both the curves of development based on longitudinal data and the curves of development inferred from normative data when analyzed in terms which approximate an absolute scale yield much the same picture. Such curves vary greatly from one characteristic to another, but, with few exceptions, they lead to the generalization that *growth and development are not in equal units per unit of time.* For each stable characteristic there is usually a period of relatively rapid growth as well as periods of relatively slow growth. Although it is not invariably true, the period of most rapid growth is likely to be in the early years and this is then followed by periods of less and less rapid growth. The differential rate of growth with time appears to us to be of greatest significance for attempts to influence growth and development and for further research on the many problems of understanding and predicting growth. We have referred to these matters throughout the book and will point up some of the implications at other places in this concluding chapter. However, here we note that for some characteristics there is as much quantitative growth in a single year at one period in the individual's development as there is in eight to ten years at other stages in his development. The importance of the influences which affect the growth of such characteristics is likely to be far greater in the periods of most rapid development than it is, at least quantitatively, in the periods of least rapid development.

One way of demonstrating this differential rate of growth is to summarize the half-development of selected characteristics (when the development by ages 18 to 20 is taken as the criterion level of growth). These are as shown on page 205.

It will be noted that, with the exception of school achievement, the most rapid period for the development of the above characteristics is in the first five years of life. While future research will, undoubtedly, yield far more precise and accurate developmental curves than are at

Height                                      Age 2½
General intelligence                        Age 4
Aggressiveness in males*                    Age 3
Dependence in females*                      Age 4
Intellectuality in males and females*       Age 4
General school achievement                  Grade 3

* Since we do not have an absolute scale for these characteristics, we have indicated the age at which one-half of the criterion variance can be predicted.

present available, it is our expectation that such research will demonstrate that we have underestimated the rapidity of the early development of these characteristics. Our present estimates, with the exception of height, are based on relatively crude measurements of these characteristics and these measurements are not entirely comparable from one age to another. As such, the present evidence on consistency and stability from the early years to maturity probably represents minimal estimates of the amount of growth which takes place in the early years.

However, we wish to remind the reader that not all human characteristics reveal the same pattern of development or stability we have found in selected characteristics. Some characteristics such as weight show a relatively uniform rate of growth from birth to age 20, whereas a characteristic such as vocational interest exhibits a very rapid development in the period from ages 14 to 20 and then changes very little for the next 20 years.

SOME RESTRICTIONS ON CHARACTERISTICS INVESTIGATED

Throughout this work we have used the term *human characteristic* or *behavioral characteristic*. In each chapter we have designated the characteristics being investigated such as height, weight, strength, general intelligence, specific aptitudes, general achievement, reading comprehension, vocabulary, sociometric status, aggression, dependence, etc.

It will be noted that in all cases a *characteristic*, as, we have used the term, refers to some measurement of human attributes or human behavior which can be represented quantitatively. Each characteristic can be represented on a quantitative scale from high to low, much to little, frequent to infrequent, large to small, etc. Although the quantitative scale does not necessarily have clearly designated units, it invariably does permit the ranking of individuals in terms of amount or presence of the characteristic.

We have included, under the term *characteristic*, attributes such as height and weight in which some dimension of the organism is selected and measured by the application of a physical scale. Such attributes require little more than the passive cooperation of the subject in order to secure a measurement. In effect, these are visible dimensions of the individual which can be calibrated with great accuracy by very simple measurement techniques.

We have also included, under the term *characteristic*, many behavioral manifestations which are usually termed traits in the field of psychological measurement. These are regularities in behaviors which are classified under a single term: vocabulary, numerical ability, aggression, dependence, memory, etc. These traits are measured by having the subject interact with particular material (problems, questions, etc.) in a standard test situation or by having an observer rate the subject after observing him under specified conditions. It should be noted that participation of the subject in some activity is usually required for the measurement.

A third way in which we have used the term *characteristic* is to refer to rather generalized qualities which are inferred from standard tests or ratings by observers. Under this heading we include strength, general intelligence, general achievement, teachers' marks, reading comprehension, sociometric status, etc. These generalized qualities are derived from an average of a number of tests or from the average of a number of ratings. We must not confuse similarity of overall rating or equality of average on such generalized qualities as representing identity, since it is possible for two individuals with very different specific ratings to have the same average test score or rating. Thus two students may have the same grade point average even though they may not have any particular grades in common.

It is clear that there are many characteristics which can be better described by the use of qualities and descriptions rather than through the use of quantities. We have not included such characteristics in this work—not because we regard them as unstable, but because we have used data analysis techniques which are more appropriate for characteristics which can be represented quantitatively.

For much the same reason we have limited ourselves to characteristics which can be represented on a single scale. This has meant that patterns of characteristics or some method of representing the individual in a multidimensional way have not been included. For many purposes, this multidimensional description is absolutely essential. However, we have limited our work here, insofar as possible, to the attempt to analyze stability and change in characteristics taken one at a time.

SOME PROPERTIES OF STABLE CHARACTERISTICS

Throughout this work, we have attempted to select and describe what we believed to be stable characteristics. As a result, the reader may be left with a somewhat biased view of human characteristics. Not all human characteristics are stable. Some characteristics may change very rapidly over time, and others may fluctuate so much that it is not possible to determine which variant is "characteristic" of the individual. In retrospect, it is possible to generalize about some of the properties of characteristics which make them stable.

Many of the stable characteristics are cumulative in that the characteristic at one point in the individual's history includes the earlier development of the same characteristic. This is clearly true of height in which the measurement at age 5 includes all the height developed to age 4, plus the increment in height made between ages 4 and 5. This also appears to be true of many of the learned characteristics such as vocabulary, reading comprehension and other school achievement. In most of these characteristics, what has been learned or developed at one point in the individual's career is apparently still present and included in the measurement at a later point.

Stable characteristics appear to either be nonreversible or at least only partially reversible. This is a corollary of the cumulative nature of stable characteristics. Whatever height has been gained to age 4 will be present at age 5 even if there is no gain in height from ages 4 to 5. It is true that there may be a loss of height during old age, but generally whatever gain has been made at one period is still retained at a later period. We suspect that many of the characteristics we find to have a high degree of stability may have partial reversibility, but we are of the opinion that generally there will be very little relative decline in a characteristic except during a period of major physiological or organic change. Many of the learned characteristics may suffer some apparent decline or loss during a period in which there is little use of the learned skills or abilities. At least, the learned material is not as readily available to the individual at a later time as it was just after the completion of the learning, for example, a foreign language, mathematics, history, etc.

Many of the stable characteristics appear to undergo relatively rapid change during an early phase of their development followed by less rapid changes. Thus height growth for boys is almost as great during the 9 months from conception to birth as it is during the 9 years from age 3 to age 12. General intelligence appears to develop as much from conception to age 4 as it does during the 14 years from age 4 to age 18.    At this point in our work, we are not entirely certain that this is a necessary condition for a stable characteristic.    However, we do believe that it is likely that a higher degree of stability will generally be found in those characteristics which develop in a negatively accelerated way than in those characteristics which develop at a fairly uniform rate.

Stable characteristics are likely to be those that have an underlying structure or that are based on underlying patterns of personality, habits, and motivation.    Thus general intelligence or general academic achievement must be manifestations of fundamental properties of the central nervous system as well as of an underlying pattern of basic habits, attitudes, and ways of relating to the world.    Such characteristics are unlikely to be altered greatly unless the underlying structure or pattern is also changed.    We would distinguish the underlying structure or pattern from the mode of expressing it.    Thus aggressiveness may arise from an underlying personality makeup, but it may be expressed in many different ways.    Girls at ages 4 to 7 may express aggressiveness in much the same ways as boys through actual physical violence, through shouting, or in some other overt way.    As girls become more "ladylike" with age they must find more socially acceptable ways of expressing aggressiveness and they must resort to more subtle modes of venting it.    The mode of expressing the underlying structure or pattern may be determined by cultural norms.

Closely related to this is the idea that a characteristic may be very stable although the expression of it may constantly change.    The problems and tasks in a general intelligence test may be very different at ages 4, 8, and 16 such that a person with an I.Q. of 100 may be expected to do more complex problems at age 16 than he was expected to do at ages 8 or 4.    It is this problem of securing comparable but different tasks at each age which makes the measurement of a stable characteristic very difficult.    A vocabulary test for first grade children will be very different from a vocabulary test for high school graduates, although vocabulary development is a moderately stable characteristic.    We have noted a number of instances in Chapter 5 in which the stability of a characteristic may be obscured by the lack of appropriate measures at different ages.

Stable characteristics are also likely to be those for which the cultural norms and values remain relatively constant.    On this basis, we might expect certain social *values* to remain very stable, whereas political *opinions* might be expected to change relatively rapidly.    General scholastic achievement may be very stable, whereas interest in art or science may shift with cultural norms and expectations. General interest in reading books and periodicals may be quite stable although the particular types of books read may vary from age to age.

That the environment does influence change in a characteristic is documented throughout this work. Studies of identical twins reared together and reared apart demonstrate that the nature of the environment will determine the extent to which individuals, with presumably identical genetic characteristics, will develop in similar or very different ways. Also studies of the effect of *changes* in environments further document the extent to which the environment influences the development of particular characteristics. This is not to say that each characteristic is equally influenced by the environment. Thus educational achievement is rather obviously influenced by environmental differences, while height is likely to be influenced by the environment to a lesser degree. Nevertheless, without taking a position on the relative influence of heredity and environment, we cannot imagine any research worker or any research in disagreement with the basic proposition that *the environment is a determiner of the extent and kind of change taking place in a particular characteristic.*

What does emerge very consistently throughout this work is the moderate to high relationship between the *magnitude* of change in a particular unit of time and the environment in which the individual is as contrasted with the relatively low degree of relationship between the change index and the initial measure on the individual. Although the environments have only been measured in a very crude way, the evidence is clear that the increments for a particular characteristic are in part determined by the environment.

For a characteristic such as height which seems so clearly to be determined by genetic factors, we conceive of the environment as enabling the individual to attain his full potential or as blunting and distorting the skeletal development. Here, the variation in environment may be seen as variation in the extent to which the individual's full potential may be achieved.

In contrast with height, which is in large part determined by heredity, is school achievement which is more clearly determined by environment. Although there must be some genetic potential for learning, the direction the learning takes is most powerfully determined by the environment. Most children in one culture will learn to read, while no child in another culture may learn to read.   Modern algebra will be learned by those who have an opportunity to learn it and it will not be learned by children who do not have this opportunity.   General school achievement, as measured by the average of a set of achievement tests or the average of the marks assigned to each student by his teachers, is likely to be greatly affected by the home, peer group, and school environments in which the children live, play, and learn.   There are clearly some environments which discourage school learning, while there are other environments which encourage and reinforce school learning.   Whatever may have been the genetic potential for learning, there is little doubt that the environment will determine what is learned and even the extent to which learning does take place.

When we turn to interests, attitudes, and personality characteristics, we are of the view that the genetic and organic base must be relatively slight, while the direction and nature of these characteristics must be largely determined by the environment in which the individual develops.

Throughout this work we have attempted to demonstrate the general curve for the development of each characteristic.   In most of these curves there appears to be a period of relatively rapid development, usually in the early years, followed by periods of less rapid development.   Much of the growth of human characteristics seems to conform to a negatively accelerated curve which may be described as a parabolic curve.   We should hasten to add that there are some characteristics which appear to grow at a relatively constant rate.   However, a most important generalization supported by our findings is that *growth is generally not in equal units per unit of time*.   Admitting, then, that there are some exceptions to the proposition, we have been interested in the varying effects of the environment on a characteristic at different phases in its development.   We would venture the proposition that *a characteristic can be more drastically affected by the environment in its most rapid period of growth than in its least rapid period of growth*. This proposition is logically supported by the rather obvious point that once a characteristic has reached its complete *development*, (height at age 20, intelligence or I.Q. at about 20, etc.) variations in the environment could have no further effect on that characteristic. Similarly, in a period of very *little development* of a characteristic, the variations in the environment could have very little effect on the characteristic.   Moving then to the more and more rapid periods of development, we would anticipate that the environment would have more and more effect on the characteristic.   Thus, to take a simple and rather extreme example, the gain in the height of boys from birth to age 3 is about 24% of mature stature, whereas the gain during the period age 3 to 6 represents about 9% of mature stature. It is likely that the environment could affect the development of stature more during the period of birth to 3 years than during the age period 3 to 6 years.

## POWERFUL ENVIRONMENTS AND CHANGE

Even the crude approximations of environments we have been able to estimate reveal sizable relationships with the measures of change. Where we have indications of extremely powerful environments, we find that the changes are relatively similar for all the individuals who are interacting with the environment. The proposition that for *powerful environments all the individuals in it will change in uniform ways* is an extreme statement which will rarely be completely supported by research findings.

Part of the difficulty in finding support for this proposition stems from the problems of measuring change. Change scores or measures are usually smaller in magnitude than initial or final measures and it is difficult to measure them reliably. Dressell and Mayhew (1954) Lord (1956, 1958), McNemar (1958), and Webster and Bereiter (1961) have pointed up the difficulty in securing reliable change measures and have suggested some statistical procedures for determining the reliability of change scores as well as procedures for increasing the reliability of change measures. Most frequently, aptitude and achievement tests are constructed in such a way that it is harder to secure significant changes on one part of the scale than on another. In a large number of studies it will be found that the changes by the students who score initially low are greater than the changes for the students who were initially high. We suspect that this unevenness of change scores may be largely attributed to the unevenness of the measuring scale. This unevenness stems from the combined effect of a ceiling on the test as well as the greater difficulty of the test items which can make the difference at the high end of the scale.

However, these are technical difficulties and there are techniques available to reduce their effect. Tests can be designed and constructed which will be more reliable and which can be used to secure reliable change scores. Tests can be designed which have much room for change at all points on the scale and which can more nearly approach equality of units. Such instruments will facilitate the development of absolute scales, such as we now have for height and weight.

In spite of these difficulties in measuring the effects of environments, we have found a few instances in which very powerful environments bring about very similar changes in the large majority of individuals. Such powerful environments represent rather extreme instances of abundance or deprivation and apparently involve most individuals in them in very similar ways.    That is, they are relatively uniform in preventing individuals from securing the necessary nutriments, learning experiences, or stimulation necessary for growth or they are so powerful in reaching all with the appropriate nutriments, experiences, and stimulation that all (or almost all) individuals are affected in similar ways and to similar extents.    In such powerful environments only relatively few individuals are able to resist the effects of the environmental pressure.

In Chapter 6 we described such extreme environments and pointed up some of the results found on individuals in these environments. Perhaps the major point to be made about such environments is their pervasiveness, that is, the individual is completely engulfed in a situation which presses him from every angle toward a particular type of development or outcome.    It is the extent to which a particular solution is overdetermined that makes for a powerful environment.


ENVIRONMENT AND STABILITY

We have already pointed to the Overlap Hypothesis as an approximation of the average amount of development taking place in a particular characteristic by a particular age.    The Overlap Hypothesis makes no assumptions about the nature of the environment other than to assume random variations in the environments under which the individuals have lived and will live to the criterion age.    On the other hand, we have pointed to the effects of the environment on particular characteristics at selected periods of development.

If we assume that the Overlap Hypothesis is an indication of the average of the development of a group of individuals up to a particular point in time and that each individual lives in an environment which is, in part, determining his growth and development, then some combination of the two should enable us to account almost fully for the development of the individuals in the group at a later point in time. That is, if the environment for each individual is measured over a particular period, then some combination of the initial measure plus a measure of the environments during the period should enable us to account for the development of each individual up to the final measure.

The proposition, *when the environment is held constant, the relationship between measurements at any two ages approaches unity*, assumes that the change taking place between the measurements is largely a function of the environment in which each individual lives during the intervening period.

Although we do lack precise measures of the environment for each characteristic, we have been able to secure crude approximations of the environment for a number of characteristics. In several of the chapters we have been able to approximate the results hypothesized on the basis of this proposition.

We are somewhat hesitant about advancing this proposition since it may be misinterpreted as contending that the environment is the complete determiner of human characteristics. It is not! Genetic factors must be taken into consideration in several ways. The genetic characteristics of an individual are included in the earliest measure of the characteristics, especially if this measure is made before very much of the environment has had an opportunity to affect the individual. The genetic influences are also at work in determining the magnitude of the average change taking place between the two measurements. That is, when the environments are distributed randomly, the average change taking place is likely to be an indication of the general time-table of development built into the specie by hereditary mechanisms.

There is a third way in which the genetic factor is included. It is included in the hereditary-environmental *interaction* already included in the first measure and it will be further projected in the growth that takes place in the period under consideration. Thus, in the period under consideration the hereditary makeup of the individual will be interactive with the environment and will, in part, determine what effect the environment will have on the individual's development. However, we believe that in some of the characteristics under consideration in this book, the genetic component must have relatively small effect in contrast to the environmental component and that the change measure will reflect the environmental effects more than it will the genetic effects.

Nevertheless, we are certain that the genetic-environment interaction and the relative weighting of the two must vary considerably from characteristic to characteristic. As more research is done with longitudinal data, we believe it is likely that characteristics will be discovered which are largely unaffected by environment except in its most extreme form of deprivation. So also we expect to learn of characteristics for which the genetic basis is so slight as to be negligible.

The characteristics represented in this work, at least those which are given fullest treatment in Chapters 2 to 5, appear to be characteristics which are subject to considerable modification by the environment. It is this modification which comes through so clearly in the proposition that *when the environment is held constant the relationship between measurements at any two ages approaches unity.*

## IMPORTANCE AND INFLUENCE OF EARLY ENVIRONMENT AND EXPERIENCE

The prolongation of the period of dependency for youth in the Western cultures has undoubtedly been a factor in desensitizing parents, school workers, and behavioral scientists to the full importance of the very early environmental and experiential influences. Youth are usually required to attend school until at least 16 years of age and the majority live at home and attend school until about age 18.

Another factor which has contributed to our lack of full awareness of the enormous influence of the early environment is the limited evidence on the effects of the early environment. And, even when such evidence is available from longitudinal studies of intelligence and personality, it has most frequently been interpreted as indicating little predictive significance for early measures of these characteristics. There appears to be an implicit assumption running through the culture that change in behavior and personality can take place at any age or stage in development and that the developments at one age or stage are no more significant than those which takes place at another.

A central finding in this work is that for selected characteristics* there is a negatively accelerated curve of development which reaches its midpoint before age 5. We have reasoned that the environment would have its greatest effect on a characteristic during the period of its most rapid development.

These findings and reasoning are supported by the results of selected studies: Lee (1951) and Kirk (1958) on intelligence, Dreizen et al. (1950) and Sanders (1934) on height, Sears et al. (1957) and Baldwin et al. (1949) on selected personality characteristics. Alexander (1961) and Bernstein (1960) further support the importance of the home environment on the language achievement of students in the early years of school. Additional support for the importance of the environment on early as well as later development may be found in the studies of siblings, fraternal, and identical twins reared together and reared apart (Newman, Freeman, and Holzinger, 1937; Burt, 1958).

* See list on page 205.

Finally, the animal research of Scott and Marston (1950) and Hebb (1949) gives support to the importance of the early environment in influencing the development of selected characteristics. The evidence referred to in the foregoing as well as the evidence summarized in Chapters 2 to 6 make it very clear that the environment, and especially the early environment, has a significant effect on the development of selected characteristics.

We believe that the early environment is of crucial importance for three reasons. The first is based on the very rapid growth of selected characteristics in the early years and conceives of the variations in the early environment as so important because they shape these characteristics in their most rapid periods of formation. We have already referred in brief detail to the evidence for this.

However, another way of viewing the importance of the early environment has to do with the sequential nature of much of human development. Each characteristic is built on a base of that same characteristic at an earlier time or on the base of other characteristics which precede it in development. Hebb (1949) has pointed out the differences in activity and exploratory behavior of animals reared in very stimulating environments in contrast to those reared under very confining conditions. Such differences in initial behavior are of significance in determining the animal's activity and intelligence at later stages in its development. Erickson (1950) has described stages in the development of human beings and the ways in which the resolution of a developmental conflict at one stage will in turn affect the resolutions of subsequent developmental conflicts. The entire psychoanalytic theory and practice is based on a series of developmental stages (Freud, 1933; Freud, 1937; Horney, 1936; Sullivan, 1953) with the most crucial ones usually taking place before about age 6. The resolution of each stage has consequences for subsequent stages. Similarly, other more eclectic descriptions of development (Havighurst, 1953; Piaget, 1932; Murray, 1938; Gesell, 1945) emphasize the early years as the base for later development. All these theoretical as well as empirical descriptions of development point up the way in which the developments at one period are in part determined by the earlier developments and in turn influence and determine the nature of later developments. For each of these viewpoints, the developments that take place in the early years are crucial for all that follows.

A third reason for the crucial importance of the early environment and early experiences stems from learning theory. It is much easier to learn something new than it is to stamp out one set of learned behaviors and replace them by a new set. The effect of earlier learning on later learning is considered in most learning theories under such terms as habit, inhibition, and restructuring. Although each learning theory may explain the phenomena in different ways, most would agree that the first learning takes place more easily than a later one that is interfered with by an earlier learning. Observation of the difficulties one experiences in learning a new language after the adolescent period and the characteristic mispronunciations which tend to remain throughout life are illustrations of the same phenomena.

Several explanations for the difficulties in altering early learning and for the very powerful effects of the early learning have been advanced. Schachtel (1949) and McClelland (1951) believe that the learning which takes place before language development is so powerful because it is not accessible to conscious memory. Others, such as Dollard and Miller (1950), Mowrer (1950), and Guthrie (1935), would attribute the power of early learning to the repeated reinforcement and overlearning over time such that the early learning becomes highly stabilized. More recently, the experimental work on imprinting in animals by Hess (1959) demonstrates the tremendous power of a short learning episode at critical moments in the early history of the organism. Hess has demonstrated that ducklings at ages of 9 to 20 hours may be imprinted to react to a wooden decoy duck as a mother duck in a ten minute learning experience and that the duckling will thereafter respond to the decoy duck in preference to real mother ducks.

Although it is possible that each type of explanation is sound, especially as it applies to different learning phenomena, all three tend to confirm the tremendous power of early learning and its resistance to later alteration or extinction.

The power of early learning must still, for humans, remain largely an inference drawn from theory, from descriptive developmental studies, and from quantitative longitudinal studies. In many respects, the attempts to describe the learning process as it takes place in the first few years of life are still far from satisfactory. We know more about the early learning of experimental animals than we do about human infants. In this writer's opinion, the most vital research problems in the behavioral sciences are those centered around the effects of early learning and early environments on humans.

## LIMITS OF CHANGE

Longitudinal research yields a table of interrelationships among measurements at different ages which closely approximate the results from normative studies when the data are interpreted in terms of scales with equal units and a defined zero point. From such data we may develop theoretical curves of the development of particular characteristics which may be compared with the observed values from specific studies. Much work is still needed to develop more satisfactory theoretical curves, but the curves already considered in Chapters 2 to 5 are likely to be generally correct even though particular details may be changed as the result of future research.

One use of these theoretical as well as observed values is to determine the limits of change which may be expected in a group or in individuals from one age period to another. If we know the initial *mean of a group* and the correlation between the initial measurements and a second set of measurements, the standard error of estimate may be used to determine the limits within which the mean score of the second set of measurements should fluctuate. Similarly, if we know the initial *measurement on an individual* and the correlation between the two measurements, the standard error of estimate may be used to determine the limits within which the second measurement should fluctuate. In general, for the characteristics studied in Chapters 2 to 5, the standard errors of estimate tend to become smaller and smaller as the characteristic develops over time; in other words, as the characteristic becomes increasingly stable, the limits of probable change for the group or for the individual become smaller and smaller.

Using present theoretical and observed values, we may now begin to ask how much change is possible from one point in development to another. How much growth in reading comprehension may be expected from one grade level to another? How much change may be expected in height development from one age to another? In addition to such questions about the extent of change, we may also wish to determine what environmental conditions and what learning or therapeutic experiences bring about the maximum amount of change and what conditions and experiences bring about the minimum amount of change in either the group or the individual. The availability of theoretical and observed values also permits new experimental research designs. Instead of asking whether an experimental group was changed more than a control group or whether the change in one group was significantly different from chance or zero, we may begin to ask whether the experimental group approached the maximum amount of growth to be theoretically expected.

We may now begin to search for educational and therapeutic procedures and conditions which consistently bring about the maximum change theoretically expected. We may also seek for procedures and conditions which transcend our theoretical limits.    The identification of such educational and therapeutic procedures would, of course, raise serious questions about the validity of our present curves of development.    Are such curves the result of present environmental conditions or do they reflect the actual limits of change in particular human characteristics?    Will new social inventions and techniques bring about far more powerful ways of changing individuals than is accounted for by our theoretical curves?    These are a few of the methodological and substantive research questions raised by the present evidence from longitudinal research.

To repeat again, one basic finding in this work is that less and less change is likely in a group or in an individual as the curve of development of a characteristic reaches a virtual plateau.    Can educational and therapeutic techniques overcome this increasing limit to change? We are, at present, somewhat pessimistic about the possibility for significant change in a characteristic once a plateau has been reached in the curve of development of that characteristic.    It is possible that very powerful environmental and/or therapeutic forces may overcome and alter the most stable of characteristics—this is yet to be demonstrated.

What is quite likely is that remedial and therapeutic techniques may enable the individual to *accept* his characteristics and to have less tension, anxiety, and emotion about them.    It is also likely that an individual may be helped to express his characteristics in more socially acceptable or even in socially approved ways.    For example, although *aggressiveness* may become a stable characteristic of an individual, he (or she) may learn how to express it in less violent and more socially acceptable ways.    Furthermore, some individuals may learn how to channel this aggressiveness so as to become very productive in scientific, scholarly, or professional pursuits.    The aggressive characteristics of a juvenile delinquent may become the acceptable behavior of a soldier in combat, a policeman on dangerous duty, or a scientific worker attacking a difficult problem.

Similarly, an individual may learn to use his level of general intelligence so effectively that he can accomplish much more intellectually than do others with much higher levels of general intelligence.

Thus, although we are pessimistic about producing major changes in a characteristic after it has reached a high level of stability, we are optimistic about the possibilities of the individual being helped to learn ways of utilizing his characteristics in more effective ways, both for his own welfare and for more productive contributions to society.

SOME RESEARCH PROBLEMS

## Measurement of Characteristics

Undoubtedly, an area of psychology which has had the greatest amount of development during the past half century has been the construction and use of instruments for the measurement of individual characteristics. As soon as a human characteristic is identified, investigated, or even named, an instrument is constructed and tried out. In many instances the designation of a particular characteristic is made in terms of the instrument which workers in the field believe to be the best or most appropriate device. General intelligence is identified with the Stanford-Binet Test, specific aptitudes are identified with the Thurstone Primary Mental Abilities Tests, personality needs are identified with the Thematic Apperception Test, etc.

The construction and use of a great variety of tests have done much to shape our conceptions of individual differences and have clearly determined which types of individual variation we emphasize and which we neglect or ignore. These measures are used increasingly for purposes of research as well as for prediction and other decision purposes in education and the behavioral sciences.

As we use these instruments in longitudinal research and in the measurement of change, it becomes clear that they must be constructed with the particular uses in mind. One source of difficulty which became apparent as we attempted to analyze longitudinal data is the question of the parallelism of the measures used at different points in time. Two tests purporting to measure the same characteristic may be sufficiently different to blur the developmental picture. However, an even more difficult problem frequently emerges when the same instrument samples different characteristics at different ages. The attempt to describe the development of general intelligence as measured by the Stanford-Binet is somewhat confused by the changing nature of the measures at different ages. At least for research purposes, we must seek methods of developing truly parallel measures at different ages or, if this fails, finding statistical and other procedures for taking into consideration both the changing nature of a particular characteristic as well as the changing nature of the measurements of the characteristic.

Most of the existing measures of individual variation were constructed to measure the status of an individual at a particular time.

As we attempt to infer change by comparing two measures of status we encounter many problems. Perhaps the most difficult of these problems is the low reliability of the change indices as contrasted with the very satisfactory reliabilities of the status measures. It will be increasingly necessary to construct instruments that will yield more precise measures of change. Such instruments will have to be constructed to sample the changes in the characteristics rather than the overall status of the characteristic at the two different points in time.

A related problem has to do with the units of measurement. It would be ideal if we could learn how to represent change as well as status measurements in terms which approximate absolute scales that have equality of unit and an absolute zero. These problems have always been present in human measurement. However, approaches to these problems have been made by Thorndike (1927), Thurstone (1928), and Gardner (1947). It seems to us that a satisfactory solution to the problem of units of measurement will enable us to determine the changing nature of a particular characteristic and will enable us to appraise more adequately the effects on such a characteristic of environmental and other variables at different points in its development.

In our own work (Bloom and Peters, 1961) and in the work of Spaulding (1960), Stone (1962), Hicklin (1962), and Yates and Pidgeon (1957) it has been demonstrated that relatively subjective ratings, marks, and other techniques for making judgments have a rather high degree of stability when properly treated. Such treatment usually requires procedures for scaling the judgments so that a common scale can be approximated. In addition, adjustments must be made for the reliability and objectivity of the judgments. It is this emphasis on reliability and objectivity by the testing movement that has led to the rejection of observational and judgmental techniques unless the evidence could be secured in no other way. If we can learn to collect this type of data in more systematic ways as well as learn to treat the data so as to correct for some of the major sources of error, such evidence will become increasingly valid as criterion data and as appropriate information at different points in the individual's development. Thus teachers' grades can be processed so as to furnish better criteria of academic achievement than is ordinarily available. Similarly, teachers' grades can furnish excellent measures of general academic achievement that will predict similar criterion measures far better than will the usual test. Perhaps the major point to be made is that judgmental, descriptive, and test data are all very useful in research, prediction, and decision making when properly treated. Each type of data may be required and is important in its own right and the assumed virtues of one should not lead to the neglect and rejection of the others.

## Measurement of Environments

We have already referred to the dearth of adequate measures of environments. The lack of such measures has, in part, led to an underemphasis on the effects of environments in behavioral science research and prediction. Throughout this work, we have made use of relatively crude measures of environments in order to analyze relationships between such measures and measures of individual variations and change. We could be reasonably confident of the results only when the environments were so extreme that there was only a slight probability that there might be errors in classifying the environments.

For the most part, the present measures of environments consist of social class status, socio-economic level, and occupational and educational level of parents. These are so general that they are likely to have only moderate relationships with the more specific environments that influence the development of physical characteristics, intelligence, and general school achievement. We are in need of more precise and specific environmental measures which are likely to be related very directly to the rate and level of development for specific individual characteristics.

The environmental measures needed are ones which can be clearly related to specific individual characteristics. These measures must include aspects of the environment which theory and empirical research suggest are most likely to have some effect on the particular characteristic. In Chapter 6 we discussed some of the aspects of environments which are most clearly related to selected individual characteristics.

There are two major problems which must be attacked in the construction of environmental measures. The first has to do with the development of instruments which identify and scale the features of the environment most clearly related to the development of the characteristic. These features are likely to include the behavior of the significant individuals in the environment, the presence and use of specific rewards and punishments, the presence and clarity of models of behavior, and the availability and use of particular facilities and materials. It is likely that such features, if properly identified, will be measurable and will be of value in differentiating among environments. Such scales can also be used to determine stability and change in the environment over time. A promising attempt to develop such measures has been reported for intelligence by Wolf (1963) and for school achievement by Dave (1963).

What is missing in the preceding approach is the process by which the individual and the environment interact to produce changes in the individual. It is likely that observational evidence as related to a theory of human development and learning will be needed to understand more fully the developmental processes. This type of attack on the problem is important if one is to understand what features of the environment must be altered or affected for optimal development to take place. Such an attack should also reveal the ways in which different parts of an individual's environment may be used to effect desirable changes in the individual as well as the ways in which environments may be created which will bring about desired developments.

Thus we would distinguish between environmental measures which are based on relevant symptoms or features of the environment and a set of procedures which more searchingly summarize the interactional processes between the individual and various features of the environment. The first approach is likely to be very useful for the attack on quantitative problems and should suffice for many problems of quantitative research. The second approach should have greater diagnostic significance and should be especially useful for the further development of theories of human development and for attempts to alter environmental conditions.

## Process of Forming and Maintaining Particular Characteristics

The longitudinal data summarized and analyzed in this book represent a series of quantitative approximations to particular aspects of growth and development. Measurements of a particular defined characteristic at various time intervals enable one to plot the general shape of the growth curve and to relate this generalized picture of growth to specific environmental characteristics. We believe this to be a very useful approach in that it makes clear a quantitative picture of growth under existing conditions. However, it does not reveal in any detail exactly how a characteristic has been formed, the process by which the characteristic is maintained and reinforced, and the process by which the characteristic may be altered.

Side by side with the quantitative longitudinal picture of growth is needed a more qualitative and process-oriented picture of the same growth. It is to be hoped that case studies of individuals can be used to supplement the longitudinal data of the future and that one will be used to illuminate the other. Furthermore, it is to be hoped that the variety of existing situations in this country and throughout the world will make possible an approach to an empirical type of research that approximates experimental conditions. Such studies can, if properly related to existing theories of learning and development, reveal the processes at work in the formation and maintenance of specific characteristics.

The development of valid environmental measures will do much to facilitate as well as stimulate this type of research. However, until we understand the processes by which particular characteristics are developed, we will be limited to measurement and prediction research. How can we describe a process of development and what principles account for the development in a great variety of specific circumstances? What forces and processes keep a characteristic from changing and what forces and processes develop the characteristic further, even after periods when normally it is considered to be fully formed. Are there some periods in the life stages of an individual when a characteristic can become markedly reorganized (for example, adolescence, college years, etc.)?

We are of the opinion that much of the stability we have reported in this work is really a reflection of environmental stability. That is, the stability of a characteristic for a group of individuals may, in fact, be explained by the constancy of their environments over time? What is needed are a number of studies in situations where the environment is clearly stable over time in contrast to situations where the environment changes over the same span of time. We have reasoned that after a characteristic has become fully developed it becomes markedly stable in spite of environmental changes. More definitive studies are needed to determine whether this is true and within what limits.

Included in such research would also be the studies of very extreme environmental conditions. Studies of concentration camps, prisons, brain washing, etc. all suggest that these very powerful environments produce considerable deterioration in characteristics which are ordinarily quite stable. At the other extreme, psychotherapy and very powerful educational environments appear to produce changes in the form of further development of intelligence, personality, and attitudes when, under other conditions, these characteristics might be expected to be quite stable. Do such extreme environmental conditions produce as much change as the present research suggests? What is it about these environments that produces the change? What is the relation between these changes in a particular characteristic and the total personality makeup of the individual? In a very stimulating paper, Rogers (1956) suggested that the very principles for the development of an individual and for the improvement of mental health could be turned around and used to destroy or at least impair the mental health and effectiveness of the individual. It is likely that research, especially of a longitudinal nature, will enable us to attack many of these problems in a more definitive way.

## New Longitudinal Studies

The length of time required for longitudinal studies and the tremendous cost of such studies probably account for the relatively small number of such investigations. With a few notable exceptions, the person responsible for setting up the study is rarely available to complete it. All too frequently, someone else was employed at the end of the study to process the data and write up the results.

In addition to time and cost, the repetitive nature of the data collecting process and the extreme amount of administrative and clerical energy required to maintain contact with the subjects and keep the records in good condition are all sources of discouragement in longitudinal research.

In spite of these difficulties, it is becoming evident that properly designed longitudinal research can be used to secure definitive answers to a great many behavioral science problems. It is our hope that the present work will stimulate increased interest in longitudinal research in many new areas.

Several suggestions for improving the strategy of longitudinal research grow out of the present effort to summarize and analyze longitudinal data.

The use of cumulative records in the schools and other social agencies provides a valuable source of data for many longitudinal studies. If such data are systematically recorded, as they are in many schools, the data may be analyzed for practical educational purposes as well as for more theoretical types of problems. Improvement in these data collections would result from better plans for testing, observations, and the recording of appropriate information about the individual and his home. Since the data are collected and recorded as a necessary or desirable part of the institution's program, the additional finances and energy required are relatively small. One of the major sources of weakness in these data collections is that the process continues without major periodic modification, probably because of the absence of a plan and because a routine once established in a social institution like the schools tends to be continued. It would be very useful if conferences could be held about every five years to review present data collections and to suggest plans for future data collections and treatment. The development of several alternative plans would result in increasingly rich collections of data both for practical use as well as for more theoretical research utilization.

The general applicability of Anderson's Overlap Hypothesis to longitudinal data suggests that longitudinal research may be broken down into smaller time segments. Thus research on the periods from birth to 5, 5 to 10, 10 to 15, 15 to 20, etc., may be done simultaneously with different (but comparable) samples of subjects. This is a modification of the proposal made by Bell (1953) to combine longitudinal and cross-sectional research to determine the convergence of the two approaches. The development of shorter longitudinal studies with the possibility of following up particular samples beyond such five-year spans should make it possible to use the longitudinal research to test particular hypotheses as well as to report the research results at more frequent intervals. This approach should also permit the research workers to secure a good approximation of the results over the entire span of time covered by the combination of the shorter studies. Thus, if properly planned and executed, a twenty-year span of development may be investigated in approximately a five-year period.

Short-span longitudinal research requires a theoretical curve of expected development over the entire span covered by the combination of the smaller studies. Such theoretical curves define the periods of most rapid and least rapid growth and enable the worker to plan his research strategy and hypotheses in relation to theoretical expectations. Theoretical growth curves can be determined from normative data based on cross-sectional samples if the results can be translated into absolute scales with equal units and a defined zero point. While some progress on the construction of absolute scales for the measurement of particular characteristics has been reported in the literature, additional research will be needed to develop more satisfactory techniques of absolute scaling. It is to be hoped that mathematical statistics will provide a basis for these new scaling procedures.

Short-span longitudinal research cannot be very fruitful if it is based solely on the measurement of individuals at fixed periods of time. The environment in which the individual is developing must also be observed or measured at the same points in time. The combination of both individual and environmental measures should yield definitive tests of particular theoretical propositions as well as richer types of data for testing many hypotheses in the behavioral sciences.

## SOME MAJOR ISSUES

### Maximization Versus Optimal Growth and Development

The research on stability and change reveals that it is possible to maximize the development of a particular characteristic like height, intelligence, school achievement, etc.   This research also suggests that it is possible to retard or stunt a particular type of development under certain kinds of environmental conditions.   The utility of maximal growth and development is not entirely clear.   Is it always desirable to bring each individual to the maximum height growth or to the highest level of intelligence he is capable of attaining?   That it is *possible* to maximize growth and development becomes clear, that it is *desirable* to maximize a particular type of development is less clear.

It is likely that there is some social or personal cost for the maximization of a particular characteristic.   Undoubtedly, the overemphasis on intelligence development is likely to be accompanied by some degree of tension and stress, concern over achievement, and preoccupation with abstractions and verbal reasoning.*   Studies of overweight persons indicate a higher rate of heart disease and shorter life span.   In contrast, some of the animal studies (McCay et al., 1935) indicate that retarded growth may be accompanied by increased longevity.

Much more research is needed on the presumed costs of maximal development of particular characteristics.   At the lower end of the scale, that is, stunted stature, mental retardation, and low levels of school achievement, etc., there seems to be little doubt that efforts to overcome these deficiencies are likely to be beneficial to both the individual and to the society.   There can be little positive value for mental retardation or school failure in a modern industrial society which places great premium on verbal learning and adaptation to rapidly changing conditions.   We are likely to see increased attention in the early years to prevent or ameliorate these deficiencies.

We are also likely to find many families and home environments which place great value on the maximization of certain types of development.   This will occur in spite of increased knowledge about the consequences of the maximization of certain characteristics.

* The research of Getzels and Jackson (1962) suggests that high levels of general intelligence and school achievement may be found in individuals who are relatively low in originality and freedom of impulse expression.

The basic question that remains is whether social institutions such as nursery schools, public schools, and welfare agencies may or should develop environments which will maximize particular characteristics or whether the environments should be oriented to the development of selected characteristics to some *optimal* degree. What is optimal may depend on further research to determine the consequences of various combinations of characteristics as well as the social and personal cost of particular types of development at various periods in the normal development of such characteristics.

## Ideal Environment

As long as we accept individual variation as primarily the consequences of genetic factors, chance, and individual (or family) free will, there are no social responsibilities other than to prevent the most clearly undesirable consequences—stunted physical development, mental retardation, delinquent behavior, etc. Perhaps also there are responsibilities for counseling and guiding the individual to make the best use of his individual characteristics so as to enable him to lead a productive and satisfying life.

As the causal relations between the environment and individual development become more clearly defined, it will be difficult for individuals or social institutions to idly observe events taking their course. Undoubtedly, individual characteristics which contribute to social chaos and clearly recognized social ills will, insofar as resources and knowledge permit, be reduced or eliminated by environmental means in the early years.

Although it may not be too difficult to recognize social ills and to take steps to reduce or eliminate them, the more positive and desirable characteristics are very difficult to determine. What are the prized characteristics in a society? These have rarely been formulated and it is likely that it would be impossible to reach a concensus on a single set of desirable characteristics for all.

In our society, it is likely that agreement might be secured on characteristics which enable the individual to adapt to a rapidly changing society as well as those which enable the individual to secure some measures of happiness in his life. Perhaps a few of the following selected from the President's Commission on Higher Education (1947) would be prized by most individuals and groups in the society:

1. To develop for the regulation of one's personal and civic life a code of behavior based on ethical principles consistent with democratic ideals.

2. To understand the common phenomena in one's physical environment, to apply habits of scientific thought to both personal and civic problems, and to appreciate the implications of scientific discoveries for human welfare.

3. To understand the ideas of others and to express one's own effectively.

4. To attain a satisfactory emotional and social adjustment.

5. To understand and enjoy literature, art, music, and other cultural activities as expressions of personal and social experience, and to participate to some extent in some form of creative activity.

6. To acquire the knowledge and attitudes basic to a satisfying family life.

7. To choose a socially useful and personally satisfying vocation that will permit one to use to the full his particular interests and abilities.

8. To acquire and use the skills and habits involved in critical and onstructive thinking.

These might be restated in many ways such as Erickson's (1950) nuclear conflicts, Havighurst's (1953) developmental tasks, Maslow's (1954) self-actualization concepts, or Roger's (1951) self-realization principles. However they may be stated and whatever the special emphasis in each viewpoint, undoubtedly, there are a number of positively stated characteristics which are regarded as desirable by many authorities. Such characteristics represent some balance between the demands of the society and the needs of the individual. At their optimal level they enable the individual to be both a productive contributor to the society as well as to lead a satisfying and useful life. Such desirable characteristics can be restated in more specific and operational terms and as such can be observed and systematically measured.

If these premises (concensus on desirable characteristics, value of these characteristics to the individual and the society, and the possibility of making the characteristics operational and observable) are granted, it is possible to approach the problem of specifying the desirable or ideal environmental conditions for their attainment. This is not to declare that all individuals will attain them to the same degree—genetic factors will provide limits. However, given proper environmental conditions a large proportion of individuals should be able to attain these characteristics to some minimal degree.

Although the research completed so far does not permit a clear prescription of an ideal environment, it does suggest some of the directions for research on this problem. The evidence presented in Chapters 2 to 5 suggests that the early environment is likely to be the significant one for the development of many of these characteristics. Perhaps the first 5 to 7 years of life are the significant years for the major beginning of most of these characteristics. Longitudinal studies of these characteristics in the early years accompanied by systematic investigations of the environmental conditions represent the minimal research procedures for the identification of some of the attributes of an "ideal" environment. That these attributes can be identified is quite likely. That this means a standard environment for all individuals is highly unlikely. What can be expected is that certain types of interaction between the child and others and between the child and the world of things, ideas, and events are more likely to lead to the development of the desired characteristics, whereas other types of interaction are likely to retard or even block the development of these characteristics. There is still a great deal of scope for the ways in which these interactions will take place and there will be many adjustments and adaptations to the particular genetic factors already present in the individual.

What is being suggested here is that more desirable child-rearing environmental conditions can be identified and implemented in the twentieth century. Such environmental conditions can be approached from a behavioral science research point of view and there is every likelihood that at least the minimal conditions for healthy development can be determined. The reader, who may be appalled by a formulation of the desirable conditions for child rearing, will find that throughout history each society or culture has had clearly specified sets of child-rearing conditions and the appropriate procedures and circumstances for child rearing were taught from one generation to another. It is only in the last half century that parents have rejected traditional approaches and have been cut off from a set of procedures which assure each parent that he is utilizing the "correct methods." The sale of "how to do it" child care booklets testifies to the need parents feel for guidance and reassurance in this area. Perhaps in the future we will find the task of being a mother or father requires far more training and preparation than has been commonly recognized. How-to-do-it booklets are not sufficient training for this very complex task.

## Social Responsibility

A central thesis of this work is that change in many human characteristics becomes more and more difficult as the characteristics become more fully developed.　Although there may be some change in a particular characteristic at almost any point in the individual's history, the amount of change possible is a declining function as the characteristic becomes increasingly stabilized.

Furthermore, to produce a given amount of change (an elusive concept) requires more and more powerful environments and increased amounts of effort and attention as the characteristic becomes stabilized.　In addition, the individual not only becomes more resistant to change as the characteristic becomes stabilized but change, if it can be produced, must be made at greater emotional cost to the individual. All this is to merely repeat once again a point made throughout this work: it is less difficult for the individual and for the society to bring about a particular type of development early in the history of an individual than it is at a later point in his history.　There is an increasing level of determinism in the individual's characteristics with increasing age and this is reflected both in the increased predictability of the characteristic and in the decreased amount of change in measurements of the characteristic from one point in time to another.

What are some of the consequences of this increasing determinism and our ability to predict the development of a characteristic from a relatively early age?

The research summarized in this work reveals that human characteristics are, in part, determined by environmental forces which can be measured.　The availability of longitudinal measurements reveals the level of prediction which can be made from one age to another.　Thus Payne (1963) has demonstrated that arithmetic achievement at grade 6 can be predicted from preschool data (before age 6) with a correlation of $+.68$.　By the end of the first grade, it can be predicted with correlations of $+.85$ or higher.　Similarly, Alexander (1961) has shown that reading comprehension at grade 8 can be predicted with a correlation of $+.73$ at grade 2 and a correlation of $+.88$ at grade 4.　Results of this type have been demonstrated on different samples.　Thus these characteristics can be predicted with relatively high levels of accuracy four to six years in advance.　With increased precision in the measurements of both the individual and the environment and with increased sophistication in the data-processing procedures, the length of the time interval and the level of predictive accuracy should be increased significantly.

As the characteristics develop to the criterion age or point, it becomes more highly predictive and less amenable to change.　As the time interval increases the criterion becomes more difficult to predict, but the characteristic is more amenable to change.

In a static agrarian society it is possible that the development of a particular characteristic would be regarded as the responsibility of the individual and/or his family. It is quite likely that the relative isolation of the individual and his family would mean that others would be unaware of the way in which the individual is developing and perhaps few would be concerned about the effects of this development.

The rapidly changing character of urban life, the increasing interdependence of people, and the increasing complexity of the society make it especially difficult for individuals who have marked problems in adaptation and learning. The declining opportunities for unskilled workers and the increasing need for a highly educated population have raised new educational requirements for our youth. School dropouts and lack of interest in higher education become problems of concern to both the individual and the society.

All these matters point up the need for increased social responsibility. If school dropouts, delinquent behavior, and frustration with the educational requirements of a society can be predicted long in advance, can we sit idly by and watch the prophecies come true? If remedial actions and therapy are less effective at later stages in the individual's development, can we satisfy a social conscience by indulging in such activities when it is far too late? When the school environment is at variance with the home and peer group environment, can we find ways of reconciling these different environments?

Put briefly, the increased ability to predict long-term consequences of environmental forces and developmental characteristics places new responsibilities on the home, the school, and the society. If these responsibilities are not adequately met, society will suffer in the long run. If these responsibilities are neglected, the individual will suffer a life of continual frustration and alienation. The responsibilities are great, the tasks ahead are difficult, and only through increased understanding of the interrelations between environments and individual development will we be able to secure more adequate solutions.

## REFERENCES

Alexander, M., 1961. The relation of environment to intelligence and achievement: a longitudinal study. Unpublished Master's Thesis, Univ. of Chicago.

Baldwin, A. L., Kalhorn, J., and Breese, F. H., 1949. The appraisal of parent behavior. *Psychol. Monogr.*, **63**, No. 4, Whole No. 299.

Bell, R. Q., 1953. Convergence: An accelerated longitudinal approach. *Child Develpm.*, **24**, 145–152.

Bernstein, B., 1960. Aspects of language and learning in the genesis of the social process. *J. Child Psychol. and Psychiatry (Great Britain)*, **1**, 313–324.

Bloom, B. S., and Peters, F., 1961.   The use of academic prediction scales for counseling and selecting college entrants.   Glencoe, Illinois: Free Press.

Burt, C., 1958.   The inheritance of mental ability.   *Amer. Psychologist,* **13,** 1–15.

Dave, R. H., 1963.   The identification and measurement of environmental process variables that are related to educational achievement.   Unpublished Ph.D. Dissertation, Univ. of Chicago.

Dollard, J., and Miller, N. E., 1950.   Personality and psychotherapy.   New York: McGraw-Hill.

Dreizen, S., Currie, C., Gilley, E. J., and Spies, T. D., 1950.   The effect of milk supplements on the growth of children with nutritive failure: II. Height and weight changes.   *Growth,* **14,** 189–211.

Dressel, P. L., and Mayhew, L. B., 1954.   General education: explorations in evaluation.   Washington: American Council on Education.

Erickson, E. H., 1950.   Childhood and society.   New York: Norton.

Freud, A., 1937.   The ego and mechanisms of defense.   London: Hogarth Press.

Freud, S., 1933.   New introductory lectures on psychoanalysis.   New York: Garden City.

Gardner, E. F., 1947.   The determination of units of measurement which are consistent with inter and intra grade differences in ability.   Unpublished Ph. D. Dissertation, Harvard Univ.

Gesell, A., 1945.   The embryology of behavior.   New York: Harper.

Getzels, J. W., and Jackson, P. W., 1962.   Creativity and intelligence.   New York: Wiley.

Guthrie, E. R., 1935.   The psychology of learning.   New York: Harper.

Havighurst, R. J., 1953.   Human development and education.   New York: Longmans, Green.

Hebb, D. O., 1949.   The organization of behavior.   New York: Wiley.

Hess, E., 1959.   Imprinting.   *Science,* **130,** 133–141.

Hicklin, W. J., 1962.   A study of long range techniques for predicting patterns of scholastic behavior.   Unpublished Ph. D. Dissertation, Univ. of Chicago.

Horney, K., 1936.   The neurotic personality of our time.   New York: Norton.

Kelley, T. L., 1924.   Statistical method.   New York: Macmillan.

Kirk, S. A., 1958.   Early education of the mentally retarded.   Urbana: Univ. of Illinois.

Lee, E. S., 1951.   Negro intelligence and selective migration: A Philadelphia test of the Klineberg hypothesis.   *Am. Sociol. Rev.,* **16,** 227–233.

Lord, F. M., 1956.   The measurement of growth.   *Ed. and Psych. Meas.,* **16,** 421–437.

Lord, F. M., 1958.   Further problems in the measurement of growth.   *Ed. and Psych. Meas.,* **18,** 437–451.

Maslow, A. H., 1954.   Motivation and personality.   New York: Harper.

McCay, C. M., Growell, M. F., and Maynard, L. A., 1935.   The effect of retarded growth upon the length of life span and upon the ultimate body size.   *J. Nutrition,* **10,** 63–70.

McClelland, D. C. et al., 1951.   Personality.   New York: William Sloane Associates.

McNemar, Q., 1958.   On growth measurements.   *Ed. and Psych. Meas.,* **18,** 47–55.

Mowrer, O. H., 1950.   Learning theory and personality dynamics.   New York: Ronald Press.

Murray, H., 1938.   Explorations in personality.   New York: Oxford Univ. Press.

Newman, H. H., Freeman, F. N., and Holzinger, K. J., 1937.   Twins: a study of heredity and environment.   Chicago: Univ. of Chicago Press.

Payne, Arlene, 1963. The selection and treatment of data for certain curriculum decision problems: a methodological study. Unpublished Ph. D. Dissertation, Univ. of Chicago.

Piaget, J., 1932. The moral judgment of the child. New York: Harcourt, Brace.

President's Commission on Higher Education, 1947. Higher education for American democracy. Washington: U. S. Government Printing Office.

Rogers, C. R., 1951. Client-centered therapy. Boston: Houghton Mifflin.

Rogers, C. R., 1956. Implications of recent advances in prediction and control of behavior. *Teachers College Record*, **57**, 5, 316–322.

Sanders, B. K., 1934. Environment and growth. Baltimore: Warwick and York.

Schachtel, E. G., 1949. "On memory and childhood amnesia," in Mullahy, P. (Ed.), A study of interpersonal relations. New York: Hermitage Press.

Scott, J. P., and Marston, M., 1950. Critical periods affecting the development of normal and maladjustive behavior of puppies. *J. Genet Psychol.*, **77**, 26–60.

Sears, R. R., Maccoby, E. E., Levin, H., 1957. Patterns of child rearing. Evanston: Row, Peterson.

Spaulding, G., 1960. Another look at the prediction of college scores. *Educational Records Bureau*. Unpublished report for the College Entrance Examination Board.

Stone, D., 1962. A methodological approach to the analysis of teacher behavior that reveals the stability of human characteristics. Unpublished Ph. D. Dissertation, Univ. of Chicago.

Sullivan, H. S., 1953. The interpersonal theory of psychiatry. New Haven: Norton.

Thorndike, E. L., 1927. The measurement of intelligence. New York: Teachers College, Columbia Univ.

Thurstone, L. L., 1928. The absolute zero in intelligence measurement. *Psychol. Rev.*, **35**, 175–197.

Webster, H. and Bereiter, C., 1961. Reliability of difference scores for the general case. Unpublished study.

Wolf, R. M., 1963. The identification and measurement of environmental process variables related to intelligence. Ph. D. Dissertation in progress, Univ. of Chicago.

Yates, A., and Pidgeon, D. A., 1957. Admission to grammar schools. London: Newnes Educational Publishing Co.

# INDEX